The Control of Oil Pollution

The Control of Oil Pollution

Edited by

J. Wardley-Smith

Formerly Technical Manager,
The International Tanker Owners Pollution Federation

Graham & Trotman
Publishers

First published in 1983 by

Graham & Trotman Limited
Sterling House
66 Wilton Road
London SW1V 1DE

ISBN 0 86010 338 2

Phototypeset in Great Britain by Input Typesetting Ltd, London
Printed in Great Britain by J. W. Arrowsmith Ltd, Bristol
Bound in Great Britain by Robert Hartnoll Ltd, Bodmin, Cornwall.

Contents

PART TWO: TREATMENT OF OIL ON THE WATER

PART THREE: TREATMENT OF OIL ON THE LAND

PART FOUR: ORGANISATIONAL ASPECTS

Preface

The first edition of this book was prepared in 1975. Since that time there has been an increase in the frequency and size of oil spills; as a consequence there has been a great deal of Research and Development on means to reduce the effects of oil pollution. The first edition was becoming increasingly out of date in a number of respects, particularly on equipment and methods.

I have been fortunate in persuading all the original authors to revise their chapters. This new edition is, therefore, an up-to-date account of the best ways, learnt from experience, to deal with pollution.

I deeply regret that, during the revision, Mr D. E. Newman died of a heart attack, so I have completed his chapter.

J. Wardley-Smith

Oxwich
June 1982

Introduction

Oil pollution of the sea has steadily increased with the increased use of oil. The recent major disaster, the explosion of the tanker *Betelgeuse* in Bantry Bay, Eire, in January 1979, which resulted in the loss of 51 lives, and the wrecking of the *Amoco Cadiz* on the coast of Brittany, with the loss, on a highly sensitive coast, of 220,000 tons of Light Arabian crude oil, have combined to bring the attention of the media of the world onto the hazards of transporting these huge quantities of oil in such vulnerable containers, while the oil well blow-out in the Gulf of Mexico (IXTOC I) emphasised the problems of under sea wells.

The Government of the United Kingdom were probably first to take action, when in 1959, as a result of incidents which occurred off the south coast of England, oil was released from a tanker being loaded in Southampton Water, and a collision off the Isle of Wight also released a considerable amount of oil. The oil from these spills, occurring in rapid succession, coated much of the shore of Southampton Water, and contamination was found to the west as far as Bournemouth, and as far east as Brighton, and thus included some of the most important holiday beaches in the British Isles. The local authorities pressed the Government of the day to do something about pollution, and in particular to advise the authorities concerned about the best way of removing the oil. As a result, the Warren Spring Laboratory, of the Department of Scientific and Industrial Research, embarked on a programme of cleaning up oil pollution—a programme which has continued, with only minor interruptions, until the present day. The wrecking of the *Torrey Canyon* on the Seven Stones Rocks, with the loss of its 117,000 tons of Kuwait crude oil, found the British not entirely unprepared, although to many, this, the largest incident to date, represents the start of oil pollution.

Before a government of a country becomes really active in drawing up measures to deal with pollution of this kind, it would seem that an incident must occur on its very shores. The United States of America, though very conscious of the problem, took little positive action until the oil well incident at Santa Barbara. The oil well blow-out IXTOC I has reinforced their concern. Until the accident to the *Juliana* off Niigata in 1970, the Japanese Government were not particularly concerned with oil pollution.

In the last fifteen years a great deal of work has been carried out on the effect of oil pollution on the marine environment, on means of preventing the escape of polluting oil and on machines and methods to deal with the oil after it has been released. Devices to do all this and discussions on their efficiency and cost have appeared in a very large range of journals, newspapers, magazines and legislation and conference reports, so that it is practically a full-time job merely to keep abreast of the constant outpouring of literature. This book has been prepared to bring together a series of chapters written by a number of individuals who are at present working actively in the field, so that anyone who has an interest in oil pollution can obtain in one place, not only a reasonably concise account of the effect of oil on the environment, but more especially an account of the various ways which are currently adopted to deal with it.

During this time, there has been a good deal of new legislation, both to reduce the amount of oil discharged and to provide penalties against those who accidentally or intentionally discharge oil, and also, in many cases, to make the spiller of the oil responsible for the clean-up. This legislation and the regulations drawn up by countries in different parts of the world with different circumstances, have in some measure prescribed the methods which can be adopted. The methods described in this book are those which can be used around the coast of the British Isles, and in many other parts of the world, but before anyone decides to adopt any of the methods suggested, they would be well advised to assure themselves they are not contravening any local legislation or bye-laws.

The book is divided into four main parts. The first deals with the source of oil which can be discharged into the water, how it can be lost or destroyed by natural causes, and the effect of this oil and of any chemicals used to disperse it, on the environment which it has polluted. This part is not intended to give a comprehensive account of the effect of oil on the environment—this would take several volumes—but rather to give the general reader some idea of the effects which could occur when oil is spilt. The references should enable any one interested in a particular aspect of the problem to start searching the voluminous literature. The second part deals with the treatment of oil on water and discusses how it can be removed, absorbed, burnt, sunk or dispersed. The third part deals with the treatment of oil on land, examines the various proposals which have been made and discusses how these methods must be adopted or modified, depending upon the actual type of land, vegetation, etc. which has been contaminated. The last brief section deals with the need to set up a satisfactory organisation to deal with oil pollution and to get it organised and tried out in detail, long before oil pollution actually arrives on the beach.

Part One

Oil and its Effects

1

Source of oil discharged into water

Oil, almost since its first formation in geological time, must have been coming to the surface of the earth and entering the water. Some of these oil seeps took place on land—the asphalt lakes of Trinidad are an example of one such enormous seep—but one wonders whether the pillar of fire by night and the column of smoke by day, which led Moses and the Israelites to the Promised Land, may not have been an oil seep burning in the desert. Such oil has not always been regarded as pollution and there are early accounts of it being collected and used for various purposes. Marco Polo, in his travels in the 13th century, refers to this as follows 'To the north lies Zorzania near the border of which there is a fountain of oil, which discharges so great a quantity of oil as to furnish loads for many camels'.[1] On the other hand, in the last hundred years, which might well be called 'The Age of Oil', escapes of oil into the environment have increased in quantity and have become more objectionable: oil pollution has arrived.

It is necessary to examine the major sources of pollution. In 1973 the world oil production was over 2,800,000,000 tons, and over the previous five years this was increasing at the rate of nearly 7.5% per year. However, the increase in oil prices reduced the use of oil and in 1975 there was a temporary fall in production. Production is increasing again and in 1979 the production was over 3000 million tons. At every stage of its transport, refining and use, oil or oil products can escape and enter the environment. Oil wells can get out of control, and release quantities of oil, either onto the land or into the sea. Tankers conveying crude oil can have accidents, or can lose part of their cargo in tank washings and operational discharges, loading and unloading can produce many oil spills, storage tanks on shore can leak, refineries discharge their effluent into rivers and the sea, and finally, the product can escape from pipelines, road vehicles or tankers in the process of transport, or from the final storage tank or the pipeline from the tank to the furnace. Even after use by burning for fuel, the resultant gases containing unburnt hydrocarbons are discharged into the air from which they can be washed by the rain back to the land, and so into the sea. Petrochemical plants discharge their effluent, containing some residual hydrocarbon products, into rivers or directly to the sea, while ordinary domestic and industrial sewage always contains waste oil. These various areas can now be briefly considered in turn, with a view to drawing up a final estimate.

Natural seeps

The natural seeps referred to above still continue and there are many places around the world where oil seeps have been noticed. Those which occur beneath the sea are a direct source of pollution. Recently, Wilson[2] has made a long and thorough examination of all records of oil seeps and has made an estimate of the total amount. He gives a range within which the true figure must lie, from a minimum of 0.2 Mt up to a maximum of 6.0 Mt per year. This is not uniformly distributed, as some 45% comes from a few high seepage areas of the world. His best estimate for an annual figure is 600,000 t.

As oil has been discharged from fissures in the sea bed in many parts of the world since geological times, one would expect the marine life in the vicinity of these seeps to be noticeably different from that in similar parts of the ocean where seeps do not occur. An examination by marine biologists of the life in the sea around Coal Oil Point, an area some 30 or 40 miles north of Santa Barbara, off the coast of California, where seeps occur, showed no distinguishable difference from that occurring in an area identical, except for the absence of seeps. Oil from these seeps, floating on the surface, is carried by the wind onto the adjacent shore, but, due to natural degradation, the amount on the shore is relatively constant and does not accumulate.

Off-shore exploration and production

It is not easy to make accurate estimates of the amount of oil entering the sea from any of the operations, particularly from drilling and exploration off-shore. The extension of exploration and production into deeper and deeper water in areas much less favourable for drilling, such as the North Slope of Alaska and the Arctic Ocean, increases the hazards and so accidental spills are more likely. Operational spills are now usually the subject of legislation and so control is much better. On a world wide basis, losses from this source must be between 100,000 and 150,000 tons per year. Estimates have been made in the United States of America of the amount of oil lost around the United States in the course of these activities, and from this, assuming that US off-shore production accounts for about 20% of off-shore oil on a world-wide basis, world losses from this source must be between 60,000 and 100,000 tons.[3] Figures of this kind are highly tentative, because a single major oil well incident can release as much oil, as minor incidents will release, in the course of a whole year.

Oil spills from oil wells

The first oil wells were on land and many people will have heard, or have seen films, of the effect known as a gusher, in which the equipment at the top of the well fails, and the oil under pressure, squirts out in a great fountain. Fortunately, this occurrence is most infrequent. Figures from the United States suggest, that some type of incident which may result in the discharge of a small amount of oil, occurs approximately once for every thousand wells drilled, while a major out-of-control situation only arises once every ten thousand wells drilled. The extension of oil fields to areas under water has increased the problems of drilling, and particularly now that oil wells are being drilled in exceedingly hazardous marine conditions such as those in Alaska and the North Sea. Because of this, public concern has been aroused both of the possibility of a spill occurring and also that, due to adverse weather and sea conditions, it might be far more difficult to stop the oil escaping. This apprehension has been recently increased by some large spills. Among them are the EKOFISK blow out in April 1977. This is an oilfield in the Norwegian sector of the North Sea, where on the Bravo platform a blow-out occurred and it is estimated that before it was stopped, 30,000 tons of oil was released. This is a small amount compared with the IXTOC I blow out. This well is in the Gulf of Mexico off Yucatan Peninsula. It spilt an estimated 4000 tons a day from June to September 1979. Earlier there had been a number of smaller, but still alarming incidents. In one spill in the Gulf of Mexico, the well went out of control, and the whole platform caught fire. In another case the oil escaped without catching fire. The method of dealing with incidents of this kind is to drill additional wells at an angle, so that they intercept the leaking well deep in the ground to block the well shaft. This is not an easy operation, but it can be done. The technique is well known and experienced crews are available. The Santa Barbara spill was rather different. The oil did not escape because of a failure of the pipe or well head, but due to the fact that the drilling passed through a section of fissured rock and the steel liners to the hole had been omitted. It is not easy to determine how much oil is actually lost in one of these cases. The flow rate of a well is known when it is working normally, but this will be very different under fault conditions. The best estimate that has been made for the world total is 150,000 tons which still remains a comparatively minor source of marine pollution. (see Table 1.6, p. 19).

In the light of the possible consequences of a major blow out in the North Sea despite all the measures to render the equipment safe, and the regulations imposed by the various States, the Oil Companies concerned have realised, that in spite of their precautions, accidents may occur. Therefore they have combined to set up an oil spill clean-up organisation, and equipment, and trained men are sited at a number of places in the area. At the same time they have formed a mutual insurance scheme called OPOL (or Offshore Pollution Liability).

In the OPOL Agreement, a sum of US $12.5 million is available to meet claims for pollution damage, which means direct loss or damage by contamination resulting from a discharge of oil. A further US $12.5 million is available to pay for clean-up operations and measures to prevent pollution.

Discharge from tankers

One of the largest sources of oil is the regular or routine discharge of oil from tankers. To appreciate how this occurs, and why, despite the present high price of oil, tankers continue to discharge and waste oil in this way, a brief account will be given of the method of operating large crude oil tankers, and also of the reasons for accidental spills.

Operation discharge

Consider a complete round trip of such a tanker, starting when it arrived at its unloading or discharge port. After unloading, a tanker is so high out of the water that its propeller might be only half covered, and the great expanse of the vessel's side can catch the wind and make the vessel very difficult to handle, particularly in confined waters. As a result, as soon as the tanks are empty, a sufficient number of them are ballasted with water, between 25 and 30% is usually enough. The tanker is then at a reasonable working level in the water and can return to the loading terminal to obtain another cargo. If nothing is done until the tanker gets to its loading point, when the ballast water is pumped out to be replaced by oil, a great deal of very oily water will be discharged at the loading terminal. This, obviously, cannot be allowed by the loading authorities and so the dirty oily water in the tanker must somehow be replaced by clean water, which can safely be discharged before a new cargo is taken.

When a tanker is unloaded, it cannot be completely clean as some oil is left on the bottom and sides of the tank and also on the cross members which are almost always present. This oil, called clingage, amounts to about 0.2% of the tank contents—the exact amount depends on the type of oil, the geometry of the tank and the temperature. In addition, some oil is left in the pipe lines of the tanker, so that the total retained, after unloading is about 0.3% of the total cargo. When the tank is ballasted with water, some of this oil mixes with the water. If the tank, which is full of water, is allowed to stand for a number of days, and if the weather is not rough enough to produce excessive motion of the ship, then most of the oil will rise to the top and form a layer. Beneath this will be a 'grey' layer of finely divided oil which rises to the surface very slowly, while in the bottom portion of the tank will be almost completely clean water. In the normal cleaning process, the whole contents of the tank are pumped over-board and then the tank is further washed by means of rotating jets of water under high pressure which can, if necessary, be heated, or perhaps have the addition of an emulsifying chemical. The whole of this oily water is then pumped over the side, the tank is clean and can be refilled with sea water so that when the vessel arrives at the loading port, the ballast can be discharged without polluting the harbour. It will be seen that the discharge of the 0.3% of the tank capacity represents 0.3% of the world volume of oil carried by sea—in total a very large amount. Consequently some twenty years ago, the oily condition of the shores along tanker routes became intolerable* and the major oil companies owning large tanker fleets decided to investigate how best the problem could be solved. They developed the method of tank cleaning commonly called load-on-top or LOT (see Fig. 1.1).

*This was despite the International Conventions limiting discharges discussed on p. 20.

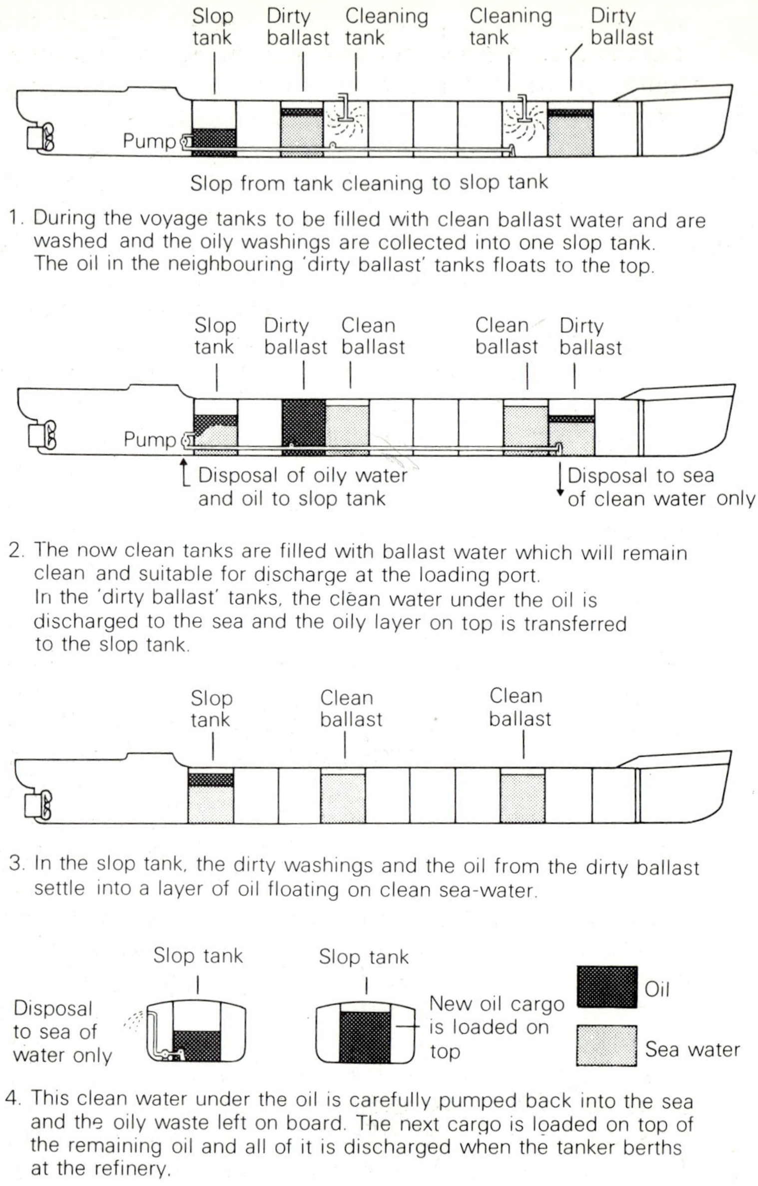

Fig. 1.1 'Load on top' system of tank cleaning

In the load-on-top process, the clean water, below the oily layer which has been allowed to settle in a tank, is pumped over-board. A careful watch is taken either visually, or by means of an oil water monitor, and the pumping is stopped when the amount of oil contained in the discharge water becomes too great. The remaining oily water is then pumped into a spare, or slop, tank. The dirty tank is then washed out as described above, but instead of discharging the wash water over-board, it too is discharged into the slop tank. The cleaned tank can then be filled with water. This process is repeated until all the tanks which were filled with ballast water at the unloading port have been treated. The tanker now has a single slop tank, which is very frequently one of the cargo tanks, containing a considerable amount of oily water. This can be allowed to settle, assisted if necessary by heating to break up any emulsion, and finally the water beneath the oil, pumped overboard as before. When the tankeı arrives at its loading port it will have a series of tanks which are oily but which contain no water, a series of tanks used for ballast which are filled with clean water and a single slop tank containing some oil and, at the bottom, a mixture of oil and water. Almost all crude oil, as it comes from the well, contains a certain amount of water so the refineries are not unaccustomed to dealing with 'wet' oil. In the load-on-top system this fact is used and the tank containing the oily water is used as a normal cargo tank and crude oil is pumped in on top. This can be dealt with by the refinery when the tanker returns with its cargo at unloading point. It must be admitted that, in this process, oil is still discharged into the sea. It is a matter of judgment to determine when the amount of oil in the water being discharged has reached an objectionable level, and even the 'clean' water contains a few parts per million of oil, perhaps as high as fifteen parts per million. Instrumentation to measure the oil content of the discharged water is being developed to comply with the 1973 Marine Pollution Convention; this will remove the human element. Nevertheless this method, even without measurements, has made an enormous reduction in the amount of polluting oil discharged by tankers, particularly from the larger ones. It is estimated that some 85% of the world's oil is now carried by tankers which have adopted the load-on-top system. These discharge under 100,000 tons of oil from tank washing while the remaining 15% which do not use the load-on-top system discharge over 500,000 tons of polluting oil. There are, of course, some circumstances in which the load-on-top system cannot be used. If the distance between the loading and discharging points is too short then there may not be sufficient time for the oil and water to separate adequately or if the voyage is rough, then the movement of water in the ballasted tank may be sufficient to prevent the oil and water separating.

For short journeys there is only one alternative, and that is for the tankers at the loading point to discharge their ballast water, not by pumping it into the sea, but by pumping it back to the shore. On shore, it can be treated in an oil/water separator and the comparatively clean water discharged to the sea. This would appear to be a most satisfactory method and it might be wondered why it is not universally adopted. However, there are a number of disadvantages. There is the extra installation of the oil/water separator on land, together with the need to run separate piping down the jetty and out to the ship. The time taken to discharge the ballast water in this way can be fairly long and may delay the loading operation. There is, however, a much more important disadvantage at many places and that is where, due to the shallow sea close to the shore, the oil is not loaded into the tanker from a jetty, but from a pipe under the sea, run out until the water is at sufficient depth for large tankers. It is then connected to a flexible pipe which rises up through the water to a floating anchored buoy (a single buoy mooring system). The tanker which has come to load, either moors by the head to that single buoy and rotates round it as the wind changes while it is loading, or, alternatively, in some cases there are additional buoys and the tanker lies in a fixed direction moored fore and aft while the oil flows from the pipeline through a flexible floating hose into the tanker. It will be appreciated, that in both these circumstances, there is only a single pipe running from the buoy back to the land so that to discharge ballast water back to the shore would entail a complete duplication of the system, and if the buoy was a long way from land, the pumping rate of discharge of ballast water using the ship's pumps would be slow and the total time would be very considerable.

A later development has been crude oil washing. This method, as its name suggests, uses part of the crude oil cargo to wash the tanks and so redisperse the residues remaining in the tanks. It can only be carried out when the tanker is fitted with an inert gas system which ensures that a flammable atmosphere can not exist in the cargo tanks. This method uses flue gas, drawn from the funnel, to replace the air above the oil in the cargo tanks. This system was developed as a result of a number of explosions which occurred when very large crude oil carriers (VLCC's) were washing their cargo tanks. In the crude oil washing system part of the cargo is used in the tank washing machines to remove clingage and residues. As this process must be carried out while discharging is taking place, the residues are discharged ashore. Tanks, cleaned in this way, have very little remaining oil, resembling tanks which have previously contained a refined oil, such as diesel fuel. Ballasting must take place as usual before sailing. The much cleaner tanks can then be cleaned, as described above, by the LOT system; but, in this case, the oil, which has to be retained on board, is only some 10% of that which would have been left from the LOT process alone.

This highly successful system has been accepted by IMCO as an alternative to the use of segregated ballast tanks (which is to have some tanks which are only used for ballast, and never for cargo, being left empty when the tanker is fully laden, this of course, reduces the carrying capacity of the tanker) which had been introduced in the 1973 IMCO (MARPOL) Convention, as amended by the 1978 Protocol.

Tanker Accidents

In general terms, the major oil producing countries are not the major users of the oil products, so that a great deal of oil has to be transported by sea. (see Fig. 1.2)

The increasing use of oil has led to an increase both in the number, and more especially, in the size of tankers. This is shown in Table 1.1 where the number of tankers in each of three size groups is given for years between 1960 and 1979. There has been over a four fold increase in tankers over 150,000dwt between 1970 and 1979 and there were not any of this size in 1965 and earlier.

With the increasing numbers of tankers on the sea more accidents have occurred, and as tankers have gone up in size so the large ones have also been involved. The first major tanker accident, that to the *Torrey Canyon*, was in 1967 when it ran on to the Seven Stones Rocks off the South West Coast of England. (see Fig. 1.3). The vessel was a complete wreck and 100,000 tons of crude oil was lost. This was the world's largest spill until twelve years later the *Amoco Cadiz* grounded off Brittany in March 1979 and lost its cargo of 220,000 tons of crude oil. Earlier in 1974 the *Metula* grounded in the Strait of Magellan while carrying 220,000 tons of crude, but in this case though some 50,000 tons were lost the remainder was salvaged and the ship safely removed (see Fig. 1.4)

It is now ten years since in the short space of six weeks in the autumn of 1969 three VLCC's suffered unexplained explosions at sea. Many investigations took place and finally reports were issued which suggested that they were due to a build up of static electricity in the cargo tanks while they were being washed. Measures to prevent this happening, in particular the inert gas system, were

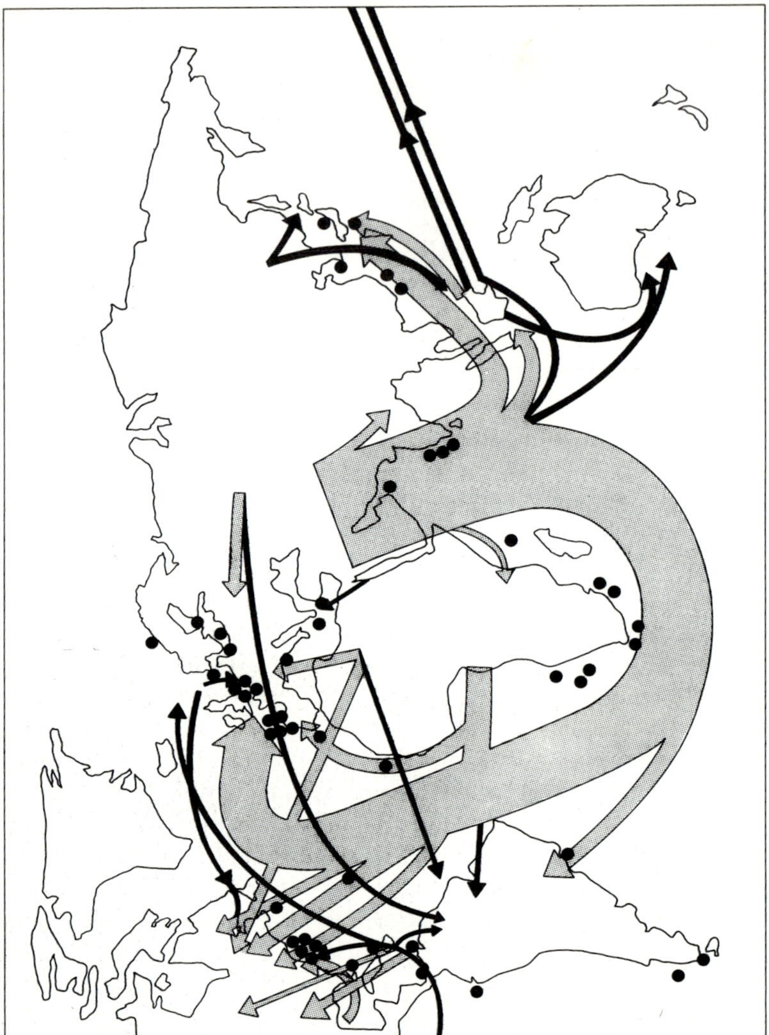

Fig. 1.2 Major oil movements and major tanker spills since 1974

TABLE 1.1
Tanker Sizes in World Fleet (numbers of ships)

As at 31 Dec.	*Below 70,000 dwt*	*70,000 to 150,000 dwt*	*Over 150,000 dwt*	*Total*
1960	2319	21	—	2340
1965	2654	128	—	2782
1970	2521	416	165	3102
1975	2449	482	634	3565
1976	2131	672	740	3543
1977	1999	684	788	3471
1978	1826	646	783	3255
1979	1770	687	744	3201

NB Tankers below 10,000 dwt and also government and non-commercial tonnage are excluded.

Fig. 1.3 *Torrey Canyon*

Fig. 1.4 *Metula* being pulled off Satellite Patch by three tugs. (This shows the eastern entrance to the Strait of Magellan.)

instigated. The losses from this cause ceased, and it appeared that major accidental spillages were being reduced. Yet in the last two years there has been a further spate of fires and explosions; many with a heavy loss of life.

In January 1979 the French tanker *Betelgeuse* exploded while alongside the Gulf Oil terminal in Bantry Bay, Eire. Fifty one people were killed, all forty three crew together with a technician who happened to be on board, and seven employees of the terminal. It has been suggested that the structure of these large tankers can be so weakened by corrosion that the extra strain put on them during the loading, discharging and ballasting operations may be enough to fracture the hull with the resulting high probability of fire and explosion. The 212,000 ton tanker *Energy Concentration* broke its back while discharging in Rotterdam in July 1980, fortunately with no fire and little pollution. This seems to be a similar case.

The amount of pollution on the shore which occurs in accidents of this kind depends on local circumstances, such as tide, wind etc. In the case of the *Silver Castle* loaded with 18,000 tons of crude oil colliding with a cargo vessel off Port Elizabeth, South Africa in April 1972, it caught fire and nineteen crew were killed. However, the fire was extinguished and the remaining 14,460 tons of oil salvaged. In contrast the *Jakob Maersk* with 83,000 tons of light Iranian crude grounded while attempting to berth at Oporto, Portugal. An explosion and fire resulted and seven crew men died. The vessel broke in two and all except the oil in one fore tank (17,000 tons) was burnt. Pollution on the shore which was only 15 km away was, considering the amount of oil lost, very small and confined to the immediate zone.

Officers and crew of tankers are human and fallible and therefore accidents can happen either due to the weather, failure of a piece of equipment or to an error or omission of a number of the crew or of the shore staff who are dealing with the cargo. There is a special type of accident which can be produced by reason of the cargo. Oil usually is an inflammable material and the lighter and more volatile components, if diluted with a large amount of air, can produce a highly explosive mixture. Although fortunately not frequent, explosion and fire is the greatest hazard to which tankers are subject. Some oil is spilt from tankers during major marine accidents of the type discussed above as a result of collision, grounding, etc. but by far the greatest number of spills occur during the loading and unloading operations.

Smaller Spills

Comprehensive data on accidental spillages from tankers is not readily available from official sources. However in 1974 the International Tanker Owners Pollution Federation instituted an Oil Spill Data Collection Scheme, under which owners of tankers, and their insurers submit standardised reports of all oil spillages from vessels entered on TOVALOP (see p. 21) to the Federation. Data from over 6500 accidental spillages all over the world have recently been published.[4]

TABLE 1.2
Oil Spills by Size: World
(1974 – 1979)

Size of Spill	*No.*	*%*
Trace	695	19
Less than 0.5 bbl	779	22
0.5 – 5 bbl	1138	32
5 – 50 bbl	568	16
50 – 5000 bbl	346	9
More than 5000 bbl	71	2

Source: Ref. 4

TABLE 1.3
World Oil Spills by Cause/Operation in Progress (1974 – 1979)

	Trace	*Barrels*					*Total*	*%*
		0–0.5	*0.5–5*	*5–50*	*50–5000*	*5000*		
Collision	6	5	4	12	38	21	86	3
Grounding	3	6	12	13	69	34	137	4
Ballasting/Deball.	101	107	177	49	18	1	453	14
Loading/Discharge	380	397	606	320	133	4	1840	56
Tank Cleaning	12	9	15	7	4	—	47	1
Bunkering	42	107	146	58	20	—	373	11
Bilge Pumping	22	19	32	11	1	—	85	3
Internal Transfer	14	28	27	20	2	—	91	3
Other	54	48	30	18	17	6	173	5

Source: Ref. 4

Their analysis shows:

1. Most spills are small, 89% being less than 50 barrels (bbl), 9% falling in the range 50 – 5000 (bbl) and only 2% being greater than 5000 (bbl) (Table 1.2)

2. Most spills (92%) occur during normal operations loading, discharging and ballasting, rather than from accident situations (Table 1.3)
3. 83% of all spills of over 5000 (bbl) are the result of collision or grounding (Table 1.3)
4. Most spills (74%) resulting from grounding or collision occur in port, port approaches or restricted waters (Table 1.4)
5. Most spills are from smaller vessels. Many of these are engaged in short runs from refinery to consumer, whereas larger tankers are usually in the long haul crude traffic.

TABLE 1.4
World Oil Spills in Port and At Sea by Cause/Operation in Progress (1974 – 1979)

	Spills in Port		*Spills at Sea*	
	No.	*%*	*No.*	*%*
Collision	87	2.1	27	0.7
Grounding	147	3.6	53	1.3
Ballasting/Deball.	556	13.5	37	0.9
Loading/Discharge	2120	51.5	141	3.4
Tank Cleaning	60	1.5	10	0.2
Bunkering	404	9.8	23	0.6
Bilge Pumping	95	2.3	13	0.3
Internal Transfer	95	2.3	11	0.2
Other	208	5.0	27	0.7
	3772		342	
Total		4114		

An attempt was made to try to identify the reasons for these spills. It will be appreciated that without a careful enquiry, and in some cases even with one, it is not easy to determine exactly what happened and then to apportion who or what was to blame. Such an examination[5] of a rather smaller number of incidents suggested that nearly half of them could be attributed to human error (see Table 1.5). Even if the reason is failure of a piece of equipment human error may still be the underlying cause. A 'leaking' sea valve, a common cause of pollution can be either the failure to close it fully—human error—or the valve may not operate correctly due to inadequate maintenance—a human error from someone else. It is interesting that the US Coast Guard attribute human error to 50 to 70% of all spills.

TABLE 1.5
Reason for Spill

Reason for fault	%
Equipment failure	34.4
Human error	46.0
Hull failure or defect	8.8
Offence denied	3.6
Shore fault	1.4
Other and not known	5.8

Estimates of the total amount of oil entering the marine environment

The total amount of oil which gets into the sea from all sources is the subject of many estimates. Jeffrey[6] in 1970 prepared an estimate based on the estimates made by various other authorities. These figures are given in the first column of Table 1.6. More recently members of the United States Coast Guard[7] have made a further study and their basic results are given in the second column of the table. It will be seen that the new total 3.8 million tons, greatly exceeds Jeffrey's total of 1.9 million tons though the latter includes 0.15 million tons for oil seepage. The US figure for waste oil, which is the major difference, is based on extrapolating to world figures from measurements of the oil content of American rivers. In the author's view, this gives much too high a figure. The third column in the table is the author's estimate which includes the seepage figure of 0.6 million tons discussed above. Even so the 3.6 million tons total does not include the possible additions from hydrocarbons initially discharged from motor cars, oil fired furnaces, power stations, industrial furnaces and domestic central heating. The total is undoubtedly very large but the transfer from the atmosphere by rainfall and then from land run-off into the sea has been neglected. However E. B. Cowell has made a recent estimate in a paper to the European Parliament, his figures are given in the last column. His total also includes discharges from dry-docking and from non-tankers which the other authors have omitted. There seems no doubt that at least two million tons, and probably a good deal more oil enters the sea each year.

TABLE 1.6
Sources of Oil in the Marine Environment ($\times 10^6$t)

Sources	*Jeffery's Estimate 1970*	*Charter et al. 1973*	*Author's Estimate 1975*	*E. B. Cowell's Estimate 1977*
From Undersea Sources			0.6	0.6
Natural Seeps				
Offshore Production	0.15	0.12	0.15	0.06
From Atmosphere				0.6
Atmospheric Fallout				
From Land Sources (other than Oil Refining)				0.3
Municipal Wastes				
Industrial Wastes	0.5	1.97[a]	1.3	0.15
Urban Run-off				0.40
River Run-off				1.40
From Oil Refining	0.3			0.06
Effluent from Coastal Refineries				
From Ocean Transportation				
Load on Top Tankers	0.1			0.11
Non Load on Top Tankers	0.6	1.07	1.0	0.5
Tanker Accidents	0.2	0.35[b]	0.35	0.3
Terminal Operations				0.001
Bilges and Bunkering	0.05	0.3	0.3	0.5
Dry Docking	—			0.25
Non Tanker Accidents	—			0.1
Total	1.90	3.81	3.60	5.331

[a]All land based discharges
[b] Accidental discharges

Agreements on oil pollution

A. International agreements

It was the oil from tank cleaning, before the adoption of the load-on-top system which, by affecting amenities on beaches and by killing diving sea birds, aroused public opinion to urge governments to take action. The British Government called an International Conference which resulted in the 1954 Convention of

Oil Pollution of the Sea. This conference prohibited the discharge of tank washings in large areas of the oceans, up to approximately one hundred miles from shore and, closer in, only permitted the discharge of oily water which did not contain more than 100 ppm of oil. Since that time there have been a large number of additional international meetings and now, under the administration of the United Nations organisation, there exists the Inter-Governmental Maritime Consultative Organisation, which is responsible for drawing up regulations concerning the discharge of oil and also other toxic or noxious cargoes into the marine environment. In 1973, a new international meeting was held in London which resulted in the 1973 Marine Pollution Convention. It is hoped that when this has been ratified and comes into force, the amount of oil discharged into the sea will be very considerably reduced. It will control and tighten up the existing Convention (1954 amended in 1959 and 1961) and under the new proposals the discharge of tank washings, anywhere in the world, will be prohibited and in many areas the discharge of even the smallest amount of oil will not be permitted. The new regulation is based on the concept that oil disappears in the marine environment and, therefore, if it is not discharged in too great quantities in any one place, it can be expected to vanish sufficiently quickly so that it will not contaminate a coast line or injure sea birds. The new Convention proposes that oil may only be discharged at a rate not exceeding 60 litres of oil per nautical mile which the ship traverses, and that the discharge only takes place while the ship is in motion and more than fifty miles from the shore. At the same time, the total amount of oil which can be discharged during any one voyage is limited.

The 1973 MARPOL Convention will not be in force until it has been ratified by 15 States who represent more than 50% of the gross tonnage of the world's merchant shipping. Ratification is very slow, partly because of the scope and complexity of the various Annexes, which include hazardous substances, sewage, and provision of reception facilities as well as oil pollution. With this in mind, the International Conference on Tanker Safety and Pollution Prevention, held in 1978 amended the 1973 Convention which permitted States to ratify the Convention Annex by Annex starting with the first. It is hoped that this will expedite the process. The 1978 Conference was held largely as a result of the wrecking of the *Amoco Cadiz* which focused world attention once more on the magnitude of the problem and the need for further regulations. The most significant change was the extension of the segregated ballast concept from tankers of 70,000 tons cargo capacity (given in 1973 Convention) to 20,000 tons dead weight for crude carriers and 30,000 tons dead weight for other oil carriers. Also all crude oil carriers above 40,000 dwt must have segregated ballast, crude oil working, or use 'clean ballast' which is similar to segregated ballast but an easier conversion in an existing construction.

Although these International regulations[8] are not yet ratified a number of States have introduced legislation and brought in the provisions before ratification. The increasing use of load-on-top and the current high price of oil have all helped to decrease the amount of oil discharged in the sea. This perhaps, applies only to oil spilt during normal operations, as the number of groundings, collisions, fires and explosions seem to have increased over the last two or three years and the amount of oil entering the ocean from these causes has consequently gone up. The 1979 Report of the Advisory Committee on Oil Pollution of the Sea,[9] which lists the beach cleaning operations undertaken by Local Authorities around the coasts of the British Isles, reports a decrease in general coastal oil pollution, though some areas had suffered from particular spills.

B. Industry agreements

The world's tanker owners appreciated that oil spills, no matter what the cause, were giving the oil industry a very bad public image and also were likely to encourage governments to introduce legislation which might hamper trade, without doing very much to reduce the number of oil spillages. The tanker owners, therefore, decided to combine and make arrangements so that they could pay the clean-up costs resulting from a spill in which a tanker was concerned. This would be done if the tanker was negligent, regardless of the degree of fault. This payment would be made either to the owner of the tanker who has incurred expenses by cleaning up the oil, or to the government concerned who has incurred expense in removing the oil, or by taking measures because the escaping oil presented a grave and imminent danger of damage by pollution to the coast lines of the country concerned. This Agreement which has been signed by the owners of 97% of the world's tankers, is called the Tanker Owners Voluntary Agreement concerning Liability for Oil Pollution (TOVALOP). It has now been in existence for over eleven years. It is administered by a small office in London called the International Tankers Owners Pollution Federation Ltd.

The tanker owners, like owners of other types of ships, combine into mutual aid groups (called P & I Clubs) who share their incurred costs under the heading of 'Protection and Indemnity'. These Clubs now include the oil spill clean-up costs. The maximum amount which can be paid out is based on the tanker size, being US$100 per gross registered ton of tanker with a maximum of US$18.6 million.

The major oil companies have also set up a fund known as CRISTAL which stands for Contract Regarding an Interim Supplement to Tanker Liability for Oil Pollution. In the event of the cost of a TOVALOP incident exceeding the limit given above, and if the cargo belongs to a member who has paid into the fund, then the CRISTAL fund may add to the total, up to a maximum of US$36 million.

In addition in 1969 the International Convention on Civil Liability for Oil Pollution Damage was drawn up. This was to provide uniform international rules and procedures for considering questions of liability and to provide adequate compensation for persons suffering from oil pollution. The liability is limited to some £86 per ton of the tanker's gross tonnage, and a total of some £9 million (these are expressed in terms of an international currency). Later, in 1971 a further Convention was to set up an International Fund for compensation for Oil Pollution Damage. This is to be provided by a levy on imports of oil in contracting states. It will provide supplementary compensation up to approximately US$35 million.

References

1. RUGOFF, M. (Ed.), *Travels of Marco Polo*, New English Library, London.
2. WILSON, R. D., 'Estimates of the Annual Input from Natural Marine Seepage', *Inputs, Fates and Effects of Petroleum in the Marine Environment*, Vol. 1, Ocean Affairs Board, National Academy of Sciences, Washington, 1973.
3. CORRINO, E. R., 'Industrial Wastes, Oils of Petroleum Origin', Paper prepared for SCEP, 1970.
4. Report by International Tanker Owners Pollution Federation Ltd., *Measures to Combat Oil Pollution*, Graham & Trotman, London, 1980.
5. WARDLEY-SMITH, J., 'Oil Spills from Tankers', Conference on Marine Ecology and Oil Pollution, Institute of Petroleum and Field Studies Councils, Aviemore, Scotland, April 1975.
6. JEFFERY, P. G., *Oil in the Marine Environment*, Report LR 156 (DC), Warren Spring Laboratory, Stevenage, Herts, England, December 1971.

7. CHARTER, D. B., SUTHERLAND, R. A. and PORRICELLI, J., 'Quantitative Estimate of Petroleum in the Ocean', *Inputs, Fates and Effects of Petroleum in the Marine Environment*, Vol. 1, Ocean Affairs Board, National Academy of Sciences, Washington, 1973.
8. For fuller details of legal aspects see CASINE, D. J. and GRANT, J. P., *The Impact of marine pollution*, Croom Helm, London, 1980.
9. ACOPS, 1 Cambridge Terrace, London NW1 4JL.

2

Removal by natural causes

Oil from many sources, has been entering the sea for generations. The present rate is measured in millions of tons each year, yet the oceans of the world are not covered with a sheet or even a sheen of oil, and although oil in small quantities can be found on many beaches they are not all heavily covered. It is clear that natural processes must make it disappear.

'Oil' is a word used to include many different substances with different chemical and physical properties. Basically, it includes the liquid which can be squeezed out of many different kinds of vegetable material—olive oil from the fruit of the olive tree, coconut oil from the 'meat' of the nut (copra). Oils are also obtained from fish and animal products—cod liver oil for example. In the context of oil pollution, what is commonly meant, is one or a mixture of the petroleum hydrocarbons which are obtained from underground deposits.

Within this immense range of oil characteristics which are found, a few major divisions can be made which describe the features which are of importance in the consideration of pollution.

Although 'oil' is a very general term, this book is primarily concerned with persistent oils, that is, oils which are visible and affect amenities. The 1954 Convention for the Prevention of Pollution of the Sea by Oil, defined 'oil' as meaning crude oil, fuel oil, heavy diesel oil, and lubricating oil, and 'oily' was construed accordingly. The 1973 Convention (see p. 20) now includes all petroleum oils except petrochemicals (which are included in Annex II). From the point of view of dealing with oils when spilt, their individual properties are most important as they influence both the rate of spreading and the rate by which they are removed from the water surface by natural causes and the problems which they present in dealing with them as pollutants. Therefore, the first section of this chapter deals with the mechanism of spreading of oil, its movement by wind and currents.

Crude Oils

Crude oils are complex mixtures of hydrocarbons of varying molecular weight and structure comprising three main chemical groups, paraffinic, naphthenic and aromatic. These hydrocarbons range from simple, highly volatile substances to complex waxes and asphaltic compounds which cannot be distilled. Oxygen, nitrogen, sulphur, vanadium, nickel and mineral salts etc. may all be present in various combinations.

The characteristics of many crude oils[1] will fall within the ranges shown in Table 2.1.

TABLE 2.1 The range of Properties of some crude oils.

Specific gravity 15/15°C		0.80–0.988	Sulphur	%wt	0.80–5.00
Initial boiling point	°C	30–125			
		(average about 2%)			
Kinematic viscosity 100°F	cS,	4–25	Wax	%wt	5–12
Pour point	°C	−35 to + 7.0	Asphaltenes	%wt	0.05–3.00

Petroleum products

The products derived from crude oils by refining will have chemical and physical characteristics which depend on the nature of the crudes and the various processes to which they have been subjected. It may be said that the sulphur, vanadium, wax and asphaltenes in the crude are associated with the higher boiling materials. Therefore, although they will appear in the light and middle distillates, they will in general become relatively more concentrated in the heavier products, i.e. medium and heavy fuel oils, and more particularly in the residues.

The typical characteristics which will give an indication of these various petroleum products are given in Table 2.2.

Lubricating Oils

These are highly refined oils, varying over a wide range of specific gravities and viscosities depending on their application. A wide range of additives is used in these oils, many of which are surface active compounds.

TABLE 2.2 Typical characteristics of some petroleum products

Gasolines (motor spirit)		
Specific gravity	15/15°C	0.68–0.77
Boiling range	°C	30–200
Flashpoint	−40°C	
Kerosene		
Specific gravity	15/15°C	0.78
Boiling range	°C	160–285
Kinematic viscosity	cS, 100°F	1.48
Flashpoint (Pensky Martens)	55°C	
Gas Oils		
Specific gravity	15/15°C	0.84
Boiling range	°C	180–360
Kinematic viscosity	cS, 100°F	3.30
Flashpoint (Pensky Martens)	77°C	
Fuel Oils (light, medium and heavy)		
Specific gravity	15/15°C	0.925–0.965
Kinematic viscosity	cS,100°F	49–862
Flashpoint	90°C Upwards	

There are a number of effects which occur when oil falls into water. It spreads, some evaporates, some dissolves, some oxidises, some is changed by bacteria, its density increases and it sinks to the bottom. Some or all of these processes effect all oils. These effects will now be considered.

Spreading

It is well known that oil spreads over water. The force of gravity acting downwards through the thickness of a considerable oil spill tends to spread it out sideways. This movement, which is the same as if oil was spilt on to a solid surface, is finally limited when the viscosity of the oil counter-balances the spreading force. On water, this gravity component rapidly becomes negligible, when compared with the surface tension effect; this force takes effect if there is a difference between the oil/air and oil/water interface tensions, and while this remains positive, the oil tends to spread out. But if a very pure oil, for example medicinal paraffin, is poured upon an absolutely clean water surface, then the oil will take up a lens-like form, the equilibrium conditions of which were calculated by Langmuir.[2] The thickness of the lens, depending on the difference between the surface and interfacial tensions, and on the density of the oil, can, in the practical case, be quite appreciable.

Fay[3] has given a theoretical treatment of the spreading of oil and divides the forces which act on the oil into three separate phases. These are:

1. Where the spreading rate is controlled by the difference in density between the oil and the water, and the speed of spreading controlled by inertia resistance.
2. Where the spreading rate is still controlled by gravity, but where viscous drag between the oil and the water limits the spreading rate.
3. Where the spreading is controlled by the surface tension difference between the oil and the water.

For practical purposes, it is only the third, or surface tension, type of spreading which has to be considered.

Blokker[4] in one of the earlier pieces of work on this subject, considered the dynamics of the spreading process and proposed an empirical formula based on the assumption that oil spreads in a uniform fashion and that the instantaneous rate of spreading is proportional to the slick thickness.

Blokker's formula, somewhat simplified, is:

$$D^3 - D_0^3 = \frac{24}{\pi} K(d_w - d_o) \frac{d_o}{d_w} V_o t$$

K = Blokker constant (which depends on the type of oil)
D = diameter of the oil field, metres
t = time
D_0 = D at time $t = 0$
d_o = density of oil
d_w = density of water
V_o = initial volume of the oil

The largest actual experiment on the spreading of oil and its disappearance was carried out by the Warren Spring Laboratory in 1972; this is described by Jeffery.[5] In this test 120 tons of light Arabian Crude Oil was discharged at sea and its appearance and dimensions observed for four days. The diagrams (see Fig. 2.1) showing the outline development and subsequent breakup of the oil slick, are particularly interesting because they show the difference between the overall picture and the thicker patches observed. It was found that if the major dimension of the slick was taken and the Blokker Constant calculated, the Constant varied with each measurement. Nevertheless, the range of values obtained was not extensive and the average value appeared to be a reasonable approximation for a spreading constant which could have been used throughout the whole four day period during which the slick was examined.

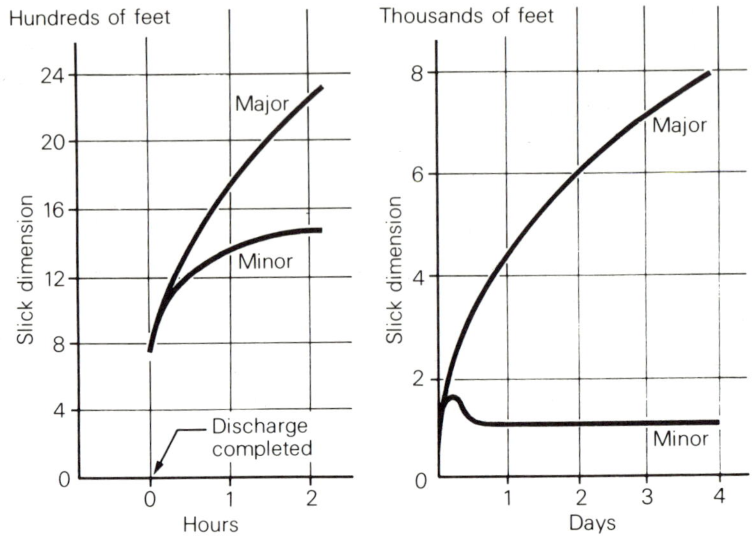

Fig. 2.1 Slick developments in major and minor dimensions with time

When crude oil is deposited on clean sea water it spreads rapidly at first. Migration into the oil/water interface of surface active components which are always contained in crude oils, can result in the formation, in 40 to 100 h, of films as thin as 4×10^{-5} inch. The surface active components will spread much more rapidly so that, as reported by Berridge in his experiments, a pronounced 'flash' of surface active components occurs.[6] These rapid flashes which are very visible from the air produce a number of peculiar reports when unskilled aerial observation is made of the large oil slicks, and gives rise to alarmist reports of 'oceans of oil six inches thick' which have appeared in the press. Observations from aircraft can indicate oil slicks many square miles in extent, while local measurements perhaps of oil blown by wind into a cove, suggest that it is inches thick. This was clearly exemplified in some of the reports about the *Torrey Canyon*. Aerial observation on the afternoon of the 24 March 1967 showed that the oil slick was 40 miles long with an average width of 10 miles. Assuming, as was reasonable, that 30,000 tons of oil had escaped from the wreck and that it was all collected in that slick, then the thickness of the oil, on average, would be one and a quarter thousandths of an inch. In actual practice the full area of the slick was probably made up of the sheen produced by the flash of the more surface active components mentioned above, while the great body of the oil would be contained in much smaller floating 'islands' discussed below, separated by the sheen. These differences can only be seen by careful aerial observation at a height not much exceeding 1000 ft.

On a uniform water surface, such as might be found on the open sea, oil should spread evenly in all directions. In practice, due to the effect of wind and waves, this is not so. The oil, quite quickly, is collected into long parallel strips or bands of oil (see Fig. 2.2). In these strips the oil is comparatively thick, while in between only 'sheen' is left. These 'winrows' seem to persist, being gradually reduced in size by evaporation etc. The tar balls found on the beach may be the final result of this process. From a practical point of view, this natural effect is the first stage of a collecting process. If a skimmer is carefully manipulated, or steered, it is possible to pass along those winrows and pick them up.

Fig. 2.2 Winrows of oil from the *Metula*

Evaporation

As the oil spreads out over water into a thin film, some of it evaporates and some of it dissolves in the water. During the initial period of weathering, there is no doubt that it is these physical processes which lead to the major loss of oil. The experiment carried out in 1967 on Kuwait crude oil, similar to that spilt by the *Torrey Canyon*, showed that under the conditions likely to be found in the English Channel at that time of the year, some 25% of the oil was lost in 24 h. The rate of evaporation loss of a given oil is undoubtedly proportional to the percentage of volatile components, that is, those with boiling points less than 270°C in the oil. These fractions also have the highest rate of solution in water, but practical experiments[7] showed that it was the medium molecular weight aromatic hydrocarbons which made up the largest proportion of dissolved components in sea water: it was found, with No. 2 fuel oil, that after 40 h in their laboratory equipment, 13% of the oil had evaporated, representing 26% of the volatile fraction, while the amount dissolved was only 0.053%. This loss correlates very well with the carbon number. Fractions containing 13 carbon atoms, or less are mostly lost in the first few days, those with up to 20 evaporating in a few weeks. This process increases the specific gravity of the remaining oil, which may ultimately finish as tar balls.

Petrochemical oxidation

Crude oil, or indeed oil of any kind, floating on the surface of the sea, is exposed to attack by atmospheric oxygen, and also to the effects of solar radiation. The extent and rapidity to which photochemical oxidation can take place depends very considerably on the physical form in which the oil is present. Oil, in the form of a thin layer or in small droplets, will be converted much more rapidly than a thicker layer of oil or a water-in-oil mousse or emulsion. The greater surface-to-volume ratio of the finely divided oil or the thin layer, allows the oxygen and the light from the sun better access to the oil. Some crude oils are coloured so that thin layers can absorb a good deal of invisible light, for example, a 0.1 mm layer of Kuwait crude oil can absorb more than 80% of the light of all wavelengths up to 667 nm, while a layer a tenth as thick absorbs 90% of ultraviolet light. Although the strength of ultraviolet light decreases with the depth of sea water, the oil is likely to be in the topmost layers and this has little effect on the rate of photochemical activity. Parker and Freegarde,[8] at the Admiralty Materials Laboratory have worked on these problems. They have pointed out that the final products can be either lighter, that is of low molecular weight or heavier than the original products. Also products of high molecular weight could be formed by radical recombination and by processes like condensation. Compounds of low molecular weight are likely to be more readily leached from an oil slick into the water, but those of a high molecular weight are likely to form the insoluble viscous gums and tars which have considerable persistance. More recently, Malins of the Convention Marine Fisheries Service has given a review of what is known about the chemical and biological conversions of petroleum products in the marine environment.[9] He suggests that some of the polar products which can be formed may be toxic to various forms of marine life. Freegarde calculates that in an 8 h day about 0.2 t/km^2 could be decomposed. This applies to slicks of 0.1 mm thickness. If the thickness is reduced, then the total rate of destruction will fall, as some of the longer wavelengths of the light will penetrate into the water beneath the oil layer.

Solution

A floating oil slick can lose oil both by the formation of a fine dispersion of oil droplets and by true solution into the sea water. It is only the lighter hydrocarbons which have any appreciable solubility so that those components which are most likely to be lost by evaporation are also those which can be leached out of the slick. The solubilities of various crude oil fractions in fresh water, derived from various references, are shown in Table 2.3.

TABLE 2.3 Solubilities of various crude oil fractions in fresh water

Fraction	*Carbon number*	*Extrapolated solubility in fresh water of corresponding n-paraffins ppm*
Kerosene	C_{10} to C_{17}	2×10^{-1} to 1×10^{-4}
Gas oil	C_{16} to C_{25}	3×10^{-4} to 10^{-8}
Lube Oil	C_{23} to C_{37}	10^{-7} to 10^{-14}
Bitumen etc.	C_{37}	less than 10^{-14}

It will be seen that the true solubility of the kerosene fractions C_{10} to C_{17} will probably be significant, but for the higher fractions, it is quite negligible. This evaporation is an important factor in preventing toxic components from entering the marine ecosystem. However, some aromatics which may give rise to long term toxicity, are less volatile and so may have a deleterious effect in the long term.

Dispersion

It will be appreciated that if the sea is rough at the time of the spill, the oil can be dispersed in very small droplets throughout a considerable depth of sea; this is shown clearly in the Report describing the *Arrow* incident.[10] In this spill, a good deal of heavy fuel oil, Bunker C, was spilt into sea water, whose temperature was just below freezing. In this case, oil was found down as far as 30 m, although the concentration was very low, in the order of 0.001 ppm. Nearer the surface, larger particles were found. A band in which such particles were found was some 10 – 25 km wide, extending along the Nova Scotia coast for about 250 km. In the cold waters of this region the rate of bacterial degradation, as discussed later, is low. Wave motion will mix the surface layer of oil into the upper layers of the sea. Some of the resulting droplets will be small and, though not a true emulsion, may still be small enough to remain for a long time in suspension, but even these amounts can be significant. For example, a suspension consisting of 1 ppm of oil distributed to a depth of one metre amounts to one ton for every one square hectometre.

The *Torrey Canyon* incident was the first time in which it had been observed and reported upon that the oil was converted by the sea into a water in oil emulsion called coloquially 'chocolate mousse' which accurately described its appearance. This emulsion did not spread so readily, was not so easily broken up by dispersants and considerably prolonged the time the oil was floating on the surface of the sea. Emulsions of this kind are of course well known industrially where they are formed when water and oil are passed through various types of machinery. They are customarily dealt with by breaking the emulsion either by heat or the addition of an appropriate emulsion breaking chemical. In the case of oil spills on the sea, neither of these alternatives would be possible and a good deal of laboratory work has been done to determine the properties of these emulsions and see how to deal with them. Berridge and his collaborators at Southampton University[11] worked on the different types of oil which could be converted into a mousse by a relatively simple mixing process. They found that a large number of crude oils, fuel and lubricating oils could be converted into a mousse, but interestingly they found it impossible to produce such an emulsion from gasolene or kerosene, whereas at the *Dona Marika* spill in Milford Haven in August 1973, where four star petrol was discharged, Baker[12] has reported the formation of mousse, however, the very vigorous sea action which occurred at the time of this wreck could not be described as 'stirring'. Jeffrey suggested that if a small amount of oil is spilt then it rapidly spreads out and the formation of a water in oil emulsion is unlikely. If, however, the quantity is greater, in this case, something over a hundred tons, even under relatively mild conditions there was a rapid and considerable pick-up of water by the oil which resulted in the formation of an emulsion, and ultimately in the production of chocolate mousse. This water in oil emulsion can contain up to seventy or eighty per cent of water. The mixture is still lower in density than the sea water but it floats nearly submerged where it usually agglomerates into a sheet of some thickness. Two or three inches is common and reports have been made of it being so thick that it was possible to stick an oar vertically in the oil without it falling over. The oil in the oil/water emulsion is not so exposed to the action of air and sunlight so that it will evaporate and polymerise more slowly. Bacterial action, as it occurs at the oil/water interface and requires both oxygen and other nutrients, is greatly slowed down. The emulsion is, therefore, much more stable in the marine environment than the original crude oil and consequently it is much more likely to be deposited on the shore.

In practice, when one is considering the hazard to an adjacent coast line resulting from a spill, the spreading characteristics of the oil will be less important than the manner in which it is moved by the wind and tidal currents. When large quantities are spilt, the resulting rafts of emulsified oil, though blown by the wind into strips or winrows, nevertheless virtually cease spreading and are moved entirely by wind and currents.

Effect of wind

Wind blowing over the sea exerts frictional forces on its surface, and it is this friction which tends to retard the wind near the surface of the sea and similarly sets the surface water in motion primarily in the same direction as the wind. This movement of the upper water surface is by friction communicated to particles of the water lower through the water, so that finally there is a current varying from a maximum on the surface to zero in contact with the sea bed. There is a further influence; this is due to the gyroscopic forces resulting from the earth's rotation. This force is at right angles to the direction of velocity of the fluid or the element of the fluid. The result of these forces will, on the whole, make wind-generated, currents-stream drift to the right of the wind in the northern hemisphere and correspondingly to the left of the wind in the southern hemisphere.

In calculating the motion due to the wind a figure for oil movement of 3.3% of the wind movement is usually taken. This is an empirical figure which, however, fits the observed movements of several large slicks on the open sea near the British Isles, and it is in the middle of the range of values suggested by various authorities. The observations following the release of oil from the *Torrey Canyon* (see Fig. 2.3) and also from the *Hamilton Trader* in Liverpool Bay, both gave answers which are close to the 3.3% of wind speed. One of the problems of calculating oil movement is that wind data is normally taken at periodic observations by neighbouring meterological stations which could in some circumstances be a considerable distance from the actual position of the slick. However, the wind speeds can be calculated from the position of the isobars on the chart.

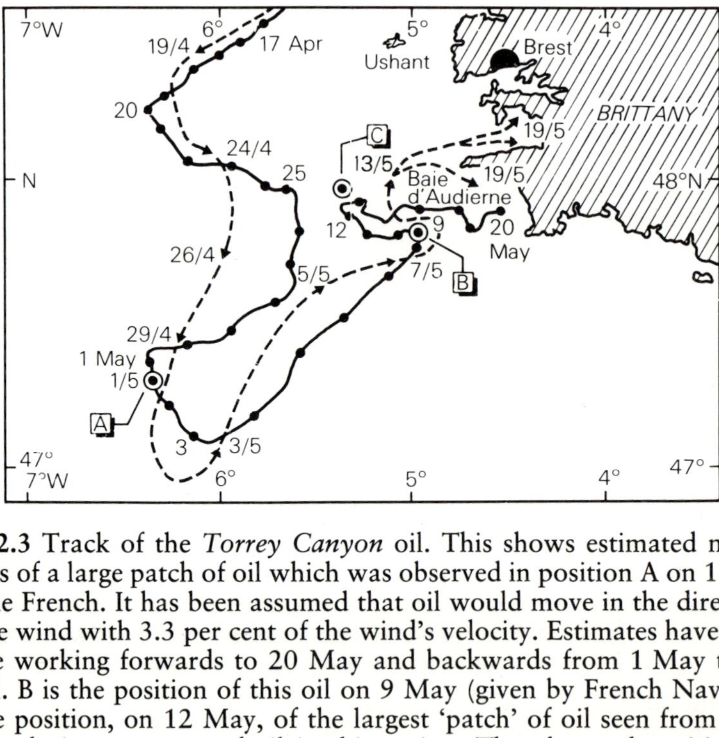

Fig. 2.3 Track of the *Torrey Canyon* oil. This shows estimated movements of a large patch of oil which was observed in position A on 1 May by the French. It has been assumed that oil would move in the direction of the wind with 3.3 per cent of the wind's velocity. Estimates have been made working forwards to 20 May and backwards from 1 May to 17 April. B is the position of this oil on 9 May (given by French Navy). C is the position, on 12 May, of the largest 'patch' of oil seen from *R.V. Sarsia* during a survey of oil in this region. The observed positions B and C are close to those predicted and the oil is shown as having come from the direction expected of oil released from the *Torrey Canyon* between 26 and 30 March. The dots on the lines mark 00.00 hours on successive days. The pecked line shows an actual track of oil movement as plotted by the French Navy and supplied to us after our own report had been completed.

Tides

In most parts of the world the sea is subject both to tides and to tidal currents, so that the actual motion of the oil has to be calculated on the basis of evaluating this factor at hourly intervals and summing these over the period of the forecast. **M** being equal to the sum of the wind factor **W**, the tidal stream vector **T**, and the residual current vector **R**, **W** as above is 3.3% of the surface wind speed and is taken parallel to the surface isobars. Data on the tidal streams in many areas of the world is readily available. The net effect of a tidal stream over a complete tide cycle is, of course, almost negligible but close to land tidal streams are important and must be included in any calculations.

Complicated calculations on the movement of oil under the action of wind, tide and currents is a complicated task so consequently various attempts have been made to put all the information in the data banks of a computer and programme it to plot the oil movement. A comprehensive system is 'Sliktrak' which is a computer simulation of offshore oil spill, clean up, effects and associated costs. It was developed by Shell[13] and has been further developed by the E & P Forum for use in the North Sea Oil Field. The computer programme applies weather data tidal movements and features of the terrain and predicts the resulting movement of the slick until the oil either disappears or reaches a coastline. The same type of programme can be adapted to deal with any oil discharge by applying appropriate factors to the source of the oil.

References

1. MARCINOWSKI, H. J., *Removal of Oil from Water Surface—* literature survey—Report No. 27/70, Stichting Concawe, The Hague, December 1970.
2. LANGMUIRE, I. J., *Chem.Phys.* **1**, 756 (1933).
3. FAY, JAMES A., 'The Spread of Oil Slicks on a Calm Sea'. *Oil on the Sea* (D. P. Hoult, Ed.), Plenum, 1969, pp. 53–63.
4. BLOKKER, P. C., 'Spreading and Evaporation of Oil Products on Water', 4th Int. Harbour Conf., Antwerp, June 1964.
5. JEFFERY, P. G., 'Large Scale Experiments on the Spreading of Oil at Sea and its Disappearance by Natural Factors, in *Conference on Prevention and Control of Oil Spills*, API and EPA, Washington, 1973, p. 469.
6. BERRIDGE, S. A., DEAN, R. A., FALLOWS, R. G. and FISH, A., 'The Properties of Persistent Oil at Sea', The Symposium of Scientific Aspects of Pollution of the Sea by Oil, Institute of Petroleum, October 1968.

7. SMITH, C. L. and MACINTYRE, W. G., 'Initial Aging of Fuel Oil Films of Sea Water', *Conference on Prevention and Control of Oil Spills*, API and EPA, Washington, 1971, p. 457.
8. *The Ultimate Fate of Crude Oil at Sea*, Int. Report No. 7. An investigation of photodecomposition of crude oil, AML Report No. AML/5/70.
9. MALINS, DONALD, C., 'The chemical and biological degradation of petroleum', *Petroleum and the Marine Environment*, Graham & Trotman, London, 1981, p. 319.
10. FORRESTER, W. D., 'Distribution of Suspended Particles following wreck of Tanker "ARROW" ', *J. Marine Research*, (May 1971).
11. BERRIDGE, S. A., *et al.*, 'Formation and stability of emulsions', *J. Inst. Pet.* 54, 333.
12. BAKER, J. M., 'Ecological Changes in Milford Haven during its history as an Oil Port', Conference on Marine Ecology and Oil Pollution, Institute of Petroleum and Field Studies Council, Aviemore, April 1975.
13. BLAIKLEY, D. R., *et al.* 'Sliktrak' a computer simulation of offshore oil spills, clean-up, effects and associated costs', Oil Spill Conference sponsored by API, EPA and USCG, 1977.

3

Bacterial degradation of crude oil

The microbiological degradation of crude oil is an important factor contributing to its disappearance at sea after preliminary physical and chemical breakdown processes have occurred. However, under adverse conditions oil may remain for decades, mainly undergraded. Even under favourable weathering conditions there may be a great persistence of spilled oil whose half life must be measured in terms of years.[1,2]

It is necessary to emphasise here that biodegradation does not always lead to the complete disappearance of the oil constituents, i.e. complete mineralisation, — the formation of carbon dioxide, water, ammonia etc. Under some conditions a variety of intermediates may accumulate and sometimes products such as waxes may be formed by unknown biosynthetic processes.[3] Migration of intermediate breakdown and the recalcitrant materials may often accumulate in bottom deposits and persist for very long periods: many if not all such residues are relatively harmless because they have low biological activities. This applies particularly to the asphalten fraction which constitutes the main residue from microbial attack on crude oils.[4]

The oil/water phase relationships and oil droplet size modify biodegradation drastically. For instance, the formation of 'chocolate mousse', water in oil emulsions, retards degradation due to slow diffusion of nutrients and oxygen to the inner layers. Tar balls are also formed at sea. These consist of varying proportions; of hydrocarbon and oxidation products associated with detritus such as sand and silt. This detritus is bound by the oil as the balls roll over the sea bed. The tar balls are sometimes beached by water movements or are carried to deeper water. Microbiological degradation occurs only on the surface layers. It is not surprising, therefore, that a surface spill of oil, or oil which is properly dispersed and weathered, is more susceptible to microbiological attack than large coherent masses.

Hydrocarbon oxidising microbes responsible for oil breakdown are relatively abundant in coastal and estuarine waters and areas subject to chronic oil pollution. Surveys by the Department of Microbiology, University College, Cardiff, in the Bristol Channel and along the west coast of England and Scotland have shown that the highest numbers of hydrocarbon oxidisers are found near the large ports and relatively very few in open, unpolluted areas. The numbers of hydrocarbon oxidising bacteria per 100 ml sea water, as estimated by MPN (most probable number) technique, were: 1 in NW Scotland coastal areas, 0–10 in the seaward areas of Bristol Channel, 10–100 in the dock areas of Barry and Swansea, with 100–1000 in the immediate vicinity of the floating crude oil used in field trials in Langstone Harbour, Portsmouth. It is thus possible that such organisms could even be used as indicators of oil pollution.

Numerous types of organisms are involved in hydrocarbon breakdown, but those more commonly isolated are Gram negative rods such as Pseudomonads and *Alcaligenes*. Other isolates include *Micrococci Corynebacteria, Mycobacteria, Nocardia, Candida* and *Penicillium*.[5]

These and other naturally occurring marine microbes normally live at the expense of low levels of hydrocarbons derived from algae and other plankton populations. On the addition of extra oil they multiply very rapidly and can reach the high levels found in oil polluted waters in a few days and may equal from 5 to 50% of the total microbial population in oil polluted coastal areas[6] where bacteria in well-oxygenated waters might oxidise oil at rates from 0.02 to 2 g/m^2 per day at 20 to 30°C.

Recently there have been suggestions that genetically engineered microbes may increase degradation.[7] The principal method for producing such so-called 'super' microbes is to insert additional material, bearing a number of genes coded for enzymes involved in the initial (proliferal) attack on oil components, carrying an extra chromosomal DNA (plasmids). There is little evidence that such organisms can make sustained improvements under natural conditions where they compete with a very mixed population.

The rates of microbial degradation, under optimum conditions where there is the potential for infection, growth and breakdown of the oil is now well established by many laboratory studies. However, useful information on the absolute rates of biodegradation of oil pollutants is more limited due to the lack of standard methods and criteria for measuring these rates.[8] However, the following results give some indication of the degree of oil breakdown under good or optimum conditions.

The average weight loss of oils from Texas, Louisiana, Oklahoma, Pennsylvania and California[9] was 45.3% of crude oil although weight loss ranged from 0.78 to 98.8%. Between 26 and 98% of the oil was degraded by mixed cultures within 30 days at 25°C. Similar work showed that there was a weight loss of 26–45%[10] but the differences in extent of weight loss of the different types of oils would seem to be related to the alkane content, oils high in alkanes being more readily degraded.[11] Approximately 50% of a Louisiana crude was utilised within 48 h by a mixed culture under conditions of vigorous agitation and aeration[12] and within 60 h microbial degradation of 35–55% of oxidisable California crude oil occurred.[13]

Crude oil is a complex mixture of a wide range of hydrocarbons with some sulphur, oxygen and nitrogen compounds. Individual components of crude oil, straight and short-branched alkanes, cycloalkanes, aromatic hydrocarbons and heterocyclic compounds are all a potential carbon and energy source for microbes and therefore biodegradable.[14, 15] It is now generally agreed, from many observations, that the aliphatic hydrocarbons are attacked more readily than the naphthenic compounds and aromatic compounds are the least susceptible to microbial attack.[16–18] In a study of the bacterial modification of petroleum it was found that a pseudomonad used paraffin oils in preference to asphaltic oil.[19] By following the degradation of Bunker C fuel oil by a mixed culture inoculum it was shown, by chromatographic analyses, that microbes selectively metabolise the *n*-alkane fraction.[20] A clear metabolic preference for *n*-alkanes was also indicated in a Louisiana crude using oil degrading microbes in an enriched sea water medium.[21] Other similar studies on Louisiana crude[22] indicated that *n*-alkanes were the most readily attacked. A preference for *n*-alkanes over isoalkanes has also been shown.[23,24]

As would be expected, no one microbial species is capable of attacking all the components of crude oil. A single species cannot even degrade all the members within a homologous series, e.g. specificity has been found within the alkanes and for aromatics such as benzene and naphthalene.[25,26] Complete degradation of oils to CO_2 and water therefore depends on the complex interplay between many microbial species. Some microbes degrade a wide range of compounds, while some are only able to utilise one hydrocarbon source for growth while simultaneously co-oxidising another hydrocarbon. Such systems of co-oxidation involve the incorporation of molecular oxygen by mono and dioxygenases[27] but only one compound is used for growth.[28]

Co-metabolism, not involving molecular oxygen, also plays an important role. Thus products formed by co-oxidation or co-metabolism by a microbe and not used for growth may be used by others and these commensal interrelationships account for a generally found phenomenon that oil fractions which appear recalcitrant when tested as growth substrates, disappear rapidly when oil is attacked by the vast range of microbes normally found in infected oils.[29]

The entry of alkanes into the microbial cell has received much discussion[30] but little experimental information is available. It is generally assumed that it takes place by direct contact. Johnson[16] suggested that the organism attaches itself to a droplet of alkane which leads to absorption into the membrane lipid.

As well as the actual chemical composition of the oil and the numbers and kinds of microbes present, there are several environmental factors which contribute to affect the rate of degradation of oil at sea.

The degradation of oil by any culture of microbes is markedly temperature dependent and has been observed over a range from about −2 to 70°C. Many factors determine this dependence not least of which is the temperature at which any culture has been isolated. Nutrient supply and oxygen tension, also affect this temperature dependance and the lack of the control of such factors have led to apparent discrepancies in the literature.[8,13,21,31–33] Most studies have been carried out at 10–30°C and with cultures isolated over this range (mesophils). From 30–50% of oil was degraded at rates of ca 130g/l/year, although laboratory experiments in Cardiff have achieved rates as high as 700g/l/year of weathered oil. There is now much interest in the rates at lower temperatures. Zobell[34] reported that even with organisms isolated at −1 to 5°C (psychrophilic) and which would be expected to degrade effectively at 2–5°C, the rates were surprisingly low, ca 3–40 g/l/year. It was suggested that this might have been due to the presence of the lighter fractions whose evaporation is slower and which inhibit the breakdown of the higher fractions.[35] Faster rates were achieved with weathered oil. Apparently faster rates were also achieved with oil and cultures taken from the Alaskan Cook Inlet where the oil (20 m/l) was almost completely degraded in 2 months but much of this has been degraded in the first week or so.[36]

Assuming an average generation time of 24 h Zobell calculated that a bacterial population of 10^8 reproducing cells per ml of mineral oil medium would consume oil for cell biomass at a rate of 1.2 mg/l/day at −1.1°C.

It is possible that low temperatures not only result in slower degradation rates but also slow the evaporation of some of the volatile components of oil and thus retard microbial growth. This was suggested,[35] as the lag periods that preceded the onset of measurable biodegradation at low temperatures were usually eliminated when 'weathered' instead of fresh crude oil was used.

Recently rapid degradation of weathered oil in treatment vessels at 45–55°C has been maintained for several months by suitably adapted populations: at 60°C and above degradation initially accelerated and subsequently fell.[37]

Agitation and aeration both accelerate the biodegradation of oils. Agitation increases the surface area of the oil droplets and also oxygen solution rates. This is important, because molecular oxygen is essential for the first step in the breakdown of many hydrocarbons by specific oxygenase enzymes. These enzymes can function at low oxygen tensions and it is likely that the oxygen content of the sea, ranging from 0 to 12 mg/l, is generally sufficiently high to allow oxygenases to operate.[38] However, the rate and extent of subsequent breakdown leading ultimately to CO_2 and H_2O will be limited by the amount of available oxygen: 3 to 4 mg of oxygen is required per mg of hydrocarbon completely oxidised to CO_2 and H_2O. Lesser quantities of oxygen are required when the hydrocarbon is only partly oxidised to intermediates, especially if these are incorporated into the cell during growth. Rivers and lakes, particularly when already polluted by sewage, often suffer from low dissolved oxygen. An attempt to get rid of an oil spill by dispersing and so increasing the rate of biodegradation could reduce the oxygen level below that necessary to maintain fish and aquatic life generally with disastrous results. If suitable electron acceptors are present many heterocyclic compounds can be completely oxidised by reactions which initially incorporate water instead of molecular oxygen.[39]

A range of oil components including alicyclics, aliphatics and aromatics are partially degraded anaerobically and a range of products accumulate including methane and fatty acids. If these accumulate they may have considerable environmental effects.

Microorganisms require available sources of C,N,P,S, Ca, K, Mg and Fe and various trace elements for growth. With the addition of a large amount of excess carbon, as oil, to the environment the amount of available nutrients, especially nitrogen and phosphorus will limit the growth and activity of oil-degrading microbes in the sea. Phosphates and nitrates in the sea fall between 1 to 100 mg/l $PO_4 - P$ and 1 to 600 mg/l $NO_3 - N$. Laboratory experiments have shown that the addition of these nutrients, as phosphates and either $NO_3 - N$ or $HN_3 - N$ in an enclosed environment, will greatly increase the rate and extent of oil degradation. Appropriate sources of nitrogen and phosphorus were used to increase the rate of degradation of polluting oil in coastal areas.[40] However, it is only in confined areas that it is thought beneficial to supplement with these nutrients; due to their high solubility in water they would rapidly be diluted out, 'seeding' areas at sea with adapted populations of crude oil. Oxidisers are unlikely to help oil breakdown where those nutrients will be limiting.

The presence of other carbon sources can affect the rate of oil degradation. High concentrations of usable organic matter such as sewage have been shown to be inhibitory.[41] The use of Armac-T treated sand for sinking oil at sea introduces organic compounds into the environment in the form of mainly long chain saturated and unsaturated primary amines. These are potential substrates for microbial growth. Results of laboratory experiments at University College, Cardiff indicate that microbes use the Armac-T as carbon and possibly a nitrogen source, causing the sunken oil to be released after a period of time.[37]

The effect of oil dispersants on the environment is not yet fully understood. Relatively non-toxic, biodegradable dispersants are used. It has been shown that biodegradation of the non-ionic oil dispersant BP1100X is rapid, when using adapted microbial populations with nitrogen and phosphorus enrichments. After 2 days incubation at 8°C and 25°C, 66% and 54%, respectively, of the surface active agent activity of the dispersant remained and after 2 weeks all the surface active agent had been degraded.[42] However, crude oil dispersed with BP1100X was shown to be only degraded rapidly in the presence of organisms adapted to both growth on the oil and the dispersant. It would appear that some of the dispersant must be degraded before the oil is made available for microbial attack. It is therefore possible that unless the oil is dispersed over a large area the freed oil droplets could coalasce before degradation takes place.

Four different chemical dispersants showed inhibitory properties on marine organisms which were more pronounced at 10°C than at 25°C. The dispersant which exhibited the poorest emulsifying property was the only one which actually promoted *n*-alkane degradation compared with the degradation of crude oil alone.[43]

In field experiments, the microbial degradation of three different types of crude oil, Kuwait, Brega and Tiajuana fresh crude oil, was followed under 'natural' conditions using an oil raft system.[42] Initially, the oils spread evenly over the surface but after only 3 weeks, the Brega and, after 8 weeks, the Kuwait crude oil had formed discret lumps of chocolate mousse. All the oils became heavily populated with bacteria during the first few months and there was a considerable increase in the number of oil-degrading microbes in the surrounding water. Highest counts obtained were 275 per ml after 2 months in the Brega tank and 1600 per ml after 5 months in the Kuwait tank. However, gas chromatography showed that even after 7 months there was still a considerable proportion of the alkane components remaining. The most significant degradation of alkanes, after this time, had occurred above carbon chain length C_{20} in the Kuwait samples. Between 9 and 12 months all the Tiajuana and Kuwait oil had formed tar balls which eventually sank due to an increase in density. (Detritus was not a contributing factor.)

Previous field experiments using oil and sinker mixtures placed on the sea bed have indicated that sunken oil is attacked by microbes but degradation is extremely slow and takes place only at the oil/water surface.[44] However, such oil gradually becomes covered by a layer of sediment and in effect 'disappears' and this may possibly be the fate of much of the oil polluting the sea.

In the first edition of this book the question was raised as to whether the natural microbial processes of oil degradation in the sea which probably extend over some 10^9 years could effectively, and if encouraged, deal with massive oil spillages. The prediction that the answer was no has been amply confirmed by more recent incidents such as *Amoco Cadiz* and smaller spillage from the *Christos Bitos*. What role has biodegradation to play? There seems no doubt that oil spillage which can be dealt with by dispersal or sinking can be left to the natural processes of microbial infection and degradation. Usually seeding is not necessary but under some circumstances nutrient addition may be advantageous especially if cheap, solid fertilisers such as basic slag and nitro chalk are employed in enclosed bays or estuaries.

With the advent of better oil/water collection and separating systems two roles for biodegradation appear worth while. The first is the treatment of dilute oil/water mixtures resulting from tanker or bunker cleaning either in dock or at sea. Several workers, including our group in Cardiff, have shown that such watery wastes can be rapidly treated to an acceptable level for legal discharge either at sea or on-shore systems.[45] In our work, flue gas was a satisfactory source of oxygen. Oil sludges separated during cleaning can also be dealt with by biological reactors which have a relatively high energy demand for mixing. The other role is for the smaller amounts of oil that constantly pollute beaches of the Atlantic and Mediterranean sea shores, due either to illegal dumping or accidents during sea transfer. These normally have to be dealt with by local authorities, especially when holiday beaches are fouled. Recent work in Cardiff and elsewhere have shown that such material is treatable[37] and that an acceptable watery effluent and solid consisting mainly of intractable and hence non-toxic residues result. It seems likely that such biotechnological processes will continue to be improved and to be employed more widely by those whose concern is the transport of this important but dwindling energy source.

References

1. BLUMER, M., EHRHARDT, M. AND JONES, J. H., 'The environmental fate of stranded crude oil, *Deep Sea Research* **20**, 239–259 (1973).
2. BLUMER, M AND SAAS, J., 'Persistance and degradation of spilled fuel oil', *Science* **197**, 1120–1122 (1972).
3. WILLIAMS, P. M., 'Organic acids in the Pacific Ocean', *Nature (Lond.)* **189**, 219 (1961).
4. EGLINTON, G., SIMONEIT, B. R. T. and ZORO, J. A., 'The Recognition of Organic Pollutants in Aquatic Sediments', *Proc. Roy Soc. B* **189**, 415 (1975).
5. FUHS, G. W., 'Der mickrobielle Abbau von Kohlrnwasserstoffen', *Arch. Mikrobiol, Abt. II.* **39**, 374–422 (1961).
6. ZOBELL, C. E., 'Microbial modification of crude oil in the sea', Proc. API and FWPCA Joint Conference on Prevention and Control of Oil Spills, New York, Dec. 1969, pp. 317–326.
7. TRIELLO, D. A., MYLROIE, J. R. and CHAKRABATY, A. M., *Use of genetically engineered multiplasmid microorganisms for the rapid degradation of fuel oil*, Applied Science, London, 1976.
8. ZOBELL, C. E., 'Microbial degradation of oil: Present status, problems and perspectives', *The Microbial Degradation of Oil Pollutants* (D. G. Ahearn and S. P. Meyers Eds) Centre for Wetland Resources. Pub. No. LSU – SG – 73 – 01, 1973, pp. 3–32.
9. ZOBELL, C. E. and PROKOP, J. F., 'Microbial oxidation of mineral oils in Barataria Bay bottom deposits', *Z. Allg. Mikrobiol* **6** 143–162 (1966).
10. DAVIS, J. B., Petroleum Microbiology, Elsevier, Amsterdam, 1967.
11. STONE, R. W., WHITE, A.G.C. and FENSKE, M. R., 'Microorganisms attacking petroleum and petroleum fractions', *J. Bact.* **39**, 91 (1940).
12. SOLI, G. and BENS, E. M., 'Bacteria which attack petroleum hydrocarbons in a saline medium', *Biotech. Bioeng.* **14** (3), 319–330 (1972).

13. MIGET, R. J., OPPENHEIMER, C. H., KATOR, H. I. and LaROCK, P. A., 'Microbial degradation of normal paraffin hydrocarbons in crude oil', Proc. API and FWPCA Joint Conference on Prevention and Control of Oil Spills, New York, Dec. 1969, pp. 327–331.
14. McKENNA, E. J. and KALLIO, R. E., 'The biology of hydrocarbons', *Ann Rev. Microbiol* **19**, 183–208 (1965).
15. ZOBELL, C. E., 'The occurrence, effects and fate of oil polluting the sea', *Adv. Wat. Pollut. Res.* **3**, 85 (1964).
16. JOHNSON, M. J., 'Utilisation of hydrocarbons by microorganisms', *Chem. Ind.* **36**, 1532 (1964).
17. ZOBELL, C. E., 'Action of microorganisms on hydrocarbons', *Bact. Rev.* **10**, 1–49 (1946).
18. DAVIS, J. B., *Petroleum Microbiology*, Elsevier, New York, 1967, 604pp.
19. DOSTALEK, M., STAUD, M. and ROSYPALOVA, A., 'The action of microorganisms on petroleum hydrocarbons', *Ceshoslov Mikrobiol* **2**, 43–47 (1956).
20. ZAJIC, J. E. and SUPPLISSON, B., 'Emulsification and degradation of "Bunker C" fuel oil by microorganisms', *Biotech. Bioeng.* **14**, 331–343 (1972).
21. KATOR, H., OPPENHEIMER, C. H. and MIGET, R. J., 'Microbial degradation of a Louisiana crude oil in closed flasks and under simulated field conditions. Proc. API/FWPCA Joint Conference on Prevention and Control of Oil Spills. American Petroleum Institute, Washington DC, 1971, pp. 287–296.
22. SOLI, G. and BENS, E. M., 'Bacteria which attack petroleum hydrocarbons in a saline medium', *Biotech. Bioeng.* **14**, 319–330 (1972).
23. ATLAS, R. M. and BARTHA, R., 'Degradation and mineralisation of petroleum by two bacteria isolated from coastal waters', *Biotech. Bioeng.* **14**, 297–308 (1972).
24. BYROM, J. A. and BEASTALL, S., 'Microbiol degradation of crude oil with particular emphasis on pollution', Microbiol, Proc. Conf., Applied Science, London, 1971, pp. 73–86.
25. BYROM, J. A., BEASTALL, S. and SCOTLAND, S., 'Bacterial degradation of crude oil, *Mar. Pollut. Bull.* **1**(2), 25–26 (1970).
26. VAN DER LINDEN, A.C. and VAN RANVENSWAAY CLAASEN, J. C., 'Hydrophobic enzymes in hydrocarbon degradation', *Lipids* **7**, (6) 437–443 (1971).
27. RAYMOND, R. L., JAMISON, V. W., and HUDSON, J. O., 'Microbial hydrocarbon co-oxidation. I. Oxidation of mono and dicyclic hydrocarbons by soil isolates of the genus Nacardia', *Appl. Microbiol.* **15**, 857–865 (1971).
28. PERRY, J. J., 'Co-oxidation involving hydrocarbons', *Microbiol. Rev.* **43**, 59 (1979).
29. HOROWITZ, GUTNICK, D. and ROSENBERG, E., 'Sequential growth of Bacteria on Crude Oil', *Applied Microbiol.* **30** (1), 10 (1975).

30. KLUG, M. J. and MARKOVETZ, A. J., 'Utilisation of aliphatic hydrocarbons by microorganisms', *Adv. Microb. Physiol* **5**, 1–43 (1971).
31. KLUG, M. J. and MARKOVETZ, A. J., 'Degradation of hydrocarbons by members of the genus Candida. I. Hydrocarbon assimilation', *Appl. Microbiol.* **15**(4), 690–693 (1967).
32. MORITA, R. Y., 'Marine psychrophic bacteria', *Oceanogr. Mar. Biol. Ann. Rev.* **4**, 105–212 (1966)
33. GUNKEL, W., 'Experimentell – okologische Untersuchungen uber die limitierenden Faktoren des mikrobiellen Olabbanes in marinen Uilieu', *Helgolaender Wiss. Meeresunters.* **15**, 210–255 (1967).
34. ZOBELL, C. E., 'Bacterial degradation of mineral oils at low temperatures', in *The Microbial Degradation of Oil Pollutants*, (D. G. Ahearn and S. P. Meyers, Eds) Centre for Wetland Resources. Pub. No. LSU – SG – 73 – 01, 1973, pp. 153–161.
35. KINNEY, P. J., BUTTON, D. K. and SCHELL, D. M., 'Kinetics of dissipation and biodegradation of crude oil Alaska's Cook Inlet', Proc. API. FWPCA Joint Conference on Prevention and Control of Oil Spills, New York, 1969, pp. 333–340.
36. ATLAS, R. M. and BARTHA, R., 'Biodegradation of petroleum in sea water at low temperatures', *Can. J. Microbiol.* **18** (12), 1851–1855 (1972).
37. BERWICK, P. G., *The Biodegradation of Oil by Hydrocarbon Oxidising Bacteria*, Ph.D Thesis, University of Wales, 1980.
38. DAVIES, J. A. and HUGHES, D. E., 'The biochemistry and microbiology of crude oil degradation', *Field Studies* **2** (suppl), 139–144 (1968).
39. QUAYLE, J. R., 'Biochemistry of hydrocarbons', In Symposium on Microbiology. London, 19–20 Sept. 1967.
40. Le PETIT, J. and BARTHELEMY, 'Les hydrocarbures en mer: Le probleme de l'epuration des zones littorales par les micro organismes', *Ann. Inst. Pasteur.* **144**. 149–158 (1968).
41. MIRONOV, O. G., 'Microorganisms growing in oil and oil products in western and central regions of the Mediterranean sea', *Revue int. Oceanogr. med.* **17**, 79–85 (1970).
42. BEASTALL, S. and HUGHES, D. E., 'Microbial aspects of the breakdown of crude oils in the environment', *J. appl. Chem. Biotechnology* **22**, 877 (1975).
43. MULLINS-PHILLIPS, G. J. and STEWART, J. E., 'Effect of four dispersants on biodegradation and growth of bacteria on crude oil' *Appl. Microbiol.* **28**, 547–552 (1974).
44. McKENZIE, P., *Degradation of Oil by Marine Microorganisms*, 1975.
45. MARSON, H. W., 'The successful treatment of Tanks in Ballast Water at the Mediterranean Site', *Prog. Wat. Techn.* **11** (No. 1), 67 (1980).
41. HUGHES, D. E. and McKENZIE, P., 'The Microbial Degradation of Oil in sea', *Proc. R. Soc. Lond. B.* **189**, 375 (1975).

4

Environmental effect of oil and the chemicals used to control it

There are many factors, acting both individually and in combination, which govern the effect of oil on marine life. These will include the type and quantity of oil, the duration and quantity of pollution experienced by the organism, the state of the oil, e.g. fresh, weathered etc., the season, with respect to the annual cycle of the organism, and the habitat and the natural stresses to which the organism is subjected. Any or all of these can affect the organism in many different ways. The effect may be lethal or sub-lethal, in the latter it can disrupt physiological or behavioural activities and death may result due to interference with feeding or reproduction. The uptake of oil can result in tainting or possible carcinogenesis either of the organism contaminated or one higher up the food chain.

The visible and well-known effects are beach pollution and the death, due to oiling, of seabirds. It is not unreasonable to wonder, if not actually to fear, whether oil might not adversely affect marine life and, in particular, harm fish—the source of much food for man and his domestic animals—either by reducing their numbers or rendering them unfit to eat.

Like all life, life in the sea depends upon solar radiation for its source of energy. Through photosynthesis, marine plants, of which by far the most important are the minute drifting organisms collectively known as phytoplankton, store this energy in a form which other organisms such as zooplankton and thc fish which prey upon it can utilise. A simplified diagram showing the broad outline of what is in reality a highly complex food web is shown in Fig. 4.1 based on Nelson Smith, 1972.[1] It is helpful to realise that at each step the efficiency of conversion of food into body weight is only about 10%. Nelson-Smith estimates that a man would require at least one kilogram of phytoplankton to gain one gram in weight. This is a useful statistic for it highlights the fact that direct effects on fish stocks eaten by man are of much greater economic importance than equivalent damage in terms of biomass lost, to the planktonic organisms on which the fish depend.

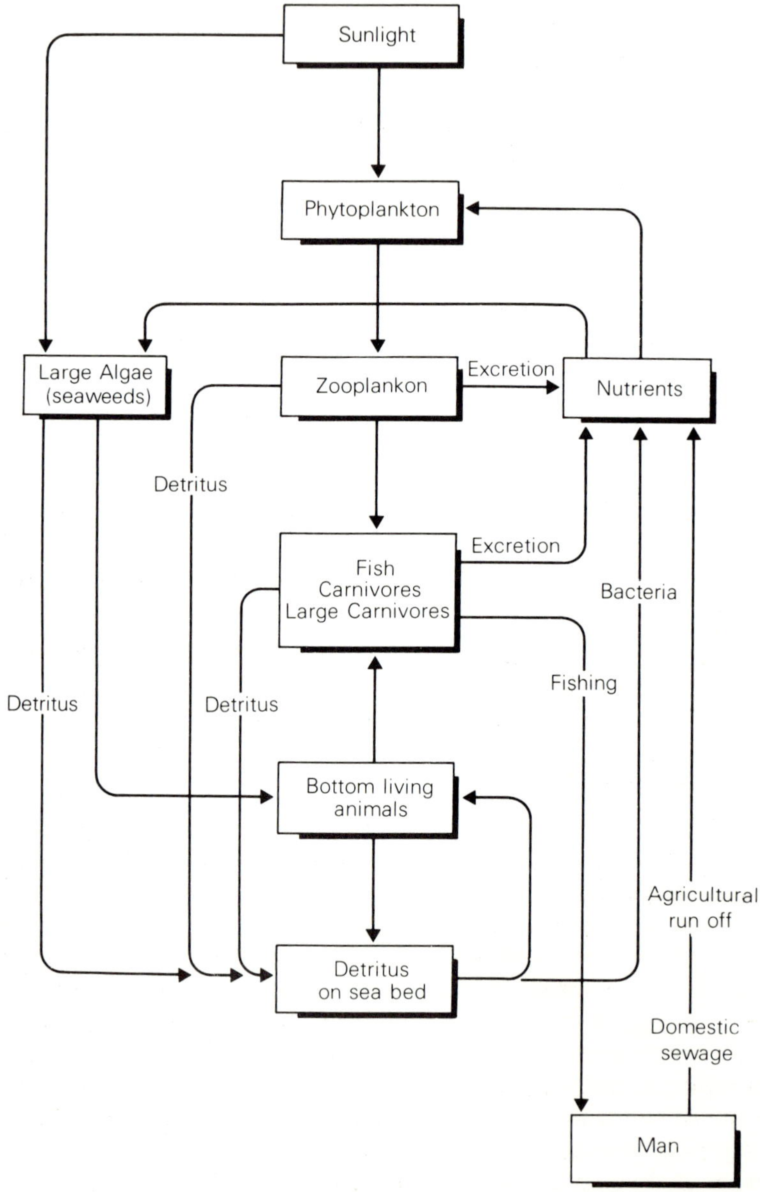

Fig. 4.1 Food cycle based on 'oil pollution and marine ecology'

The food web diagram also highlights the fact that fish, and other organisms well removed from the primary production stage represented by the phytoplankton, have, by proxy, consumed relatively enormous quantities of organic materials some of which, although only present in small concentrations in phytoplankton, may collect at the higher trophic levels of the food web. The question of whether there is an accumulation of petroleum hydrocarbons compounds, if they represent a credible threat to the resource or the consumer, and the other possible effects on marine life of oil and of chemicals and methods used to clean it up, are considered in Sections 1 – 3.

1. Plankton

At almost all major oil spills, laymen and ecologists alike express fears about possible harmful effects on plankton. Concern can usually be summarised under three headings:

(i) A layer of oil on the surface will, by occluding light, cause a serious drop in photosynthesis by phytoplankton.
(ii) The oil layer will interfere with gaseous interchange at the sea surface and dissolved oxygen levels will thereby be lowered.
(iii) Oil and dispersants will exert serious toxic effects on plankton.

How real are these effects and what is their ecological significance? The first and second are the indirect consequences of measurable physico/chemical phenomena and will be treated first.

1.1. Occlusion of light

Ottway[2] studied the light transmitting properties of 20 crude oils made up as 5% solutions in white spirit and showed that different crude oils vary widely in this respect. Oils high in asphaltenes and sulphur were found to absorb most light and, for all oils, the absorption spectra exhibited maxima at the blue end of the spectrum (325 – 350 nm), indicating that the same group of compounds is mainly responsible in all of the oils tested. A number of photosynthetic pigments found in phytoplankton also have absorption maxima at the blue end of the spectrum. Potentially, at least, their photosynthetic activity could be impeded by crude oils spilt at sea. It is important to remember the scale of the problem. A huge slick of black oil covering 2000 square miles of the North Sea actually covers less than 1% of its area. It is safe to conclude that occlusion of light by spilled oil, a phenomenon easily demonstrated in the laboratory, has no ecological significance for the plankton of the open sea.

1.2. Effects on dissolved oxygen levels

The solubility of oxygen in most hydrocarbon mixtures is similar to that in sea water. An oil slick is not, therefore, an impenetrable barrier to gaseous interchange. For a full discussion of this issue, the reader should consult Nelson-Smith;[1] suffice it to say that in a 2 day trial, Brown and Reid[3] found that boiled water, beneath a 17 mm layer of oil, absorbed 73% as much oxygen as the control. Confirmation that oil slicks do not interfere significantly with gaseous interchange at the surface of the open sea is provided by an observation by the Scripps Institution of Oceanography,[1] on dissolved oxygen levels below a heavy oil slick in the Santa Barbara Channel. The oxygen saturation beneath the slick was 98.5% of that in water outside the slick area.

1.3. Toxic effects

There is abundant evidence[1,4–6] that freshly-spilled crude oils contain low-boiling substances which are acutely toxic to planktonic organisms. The most toxic of these are water-soluble aromatic derivatives (aromatic hydrocarbons and naphthenoaromatics). Low molecular weight alkanes (10 carbon atoms or less) may induce narcosis in high concentrations but these are unlikely to result from a crude oil spill in the open sea.[6] The aromatic solvents used in early solvent-emulsifiers were also found to be highly toxic to planktonic organisms at levels of less than one part per million.[1,4,7] In later formulations of equivalent dispersing power, the use of non-aromatic solvents has reduced toxicity by over two orders of magnitude.

Useful as laboratory assessments of toxicity can be for comparing the acute toxicity of oils and dispersant chemicals to small numbers of organisms, they are rarely a reliable guide to effects on wild populations. In that such tests are necessarily short term, they under-estimate toxicity, but on the other hand, the toxicant concentrations used in laboratory experiments are rarely maintained for any length of time in the sea. Evaporative losses and the rapid dilution of residual, soluble compounds ensures that toxic effects are short-lasting, even for high toxicity, solvent-emulsifiers of the aromatic type. Thus, although highly-localised short-term effects on plankton were observed following the massive use of high aromatic solvent-emulsifiers to disperse *Torrey*

Canyon oil[4] there was no evidence that any planktonic species was significantly affected at the population level. The *Torrey Canyon* experience was quite exceptional in the amount and toxicity of the materials used to disperse the oil, and even there it would seem that the rapid evaporation and dilution of the toxicants combined with the dispersed nature and rapid turnover of the plankton to minimise the ecological impact. For other major spillages, including the notorious Santa Barbara blowout, toxic effects on the plankton have proved to be virtually undetectable.[1] The plankton of Milford Haven was examined to see if the development of a major oil port had had any effect. The report[8] stated that changes which have occurred are no greater than might have been produced by natural environmental changes and population dynamics. Sanbourn[9] suggests however that the effect on benthic larvae when they are in the upper layers of the ocean may be more long lasting as benthic animals require years to reach maturity and so the effect on the larvae may endure longer. The only planktonic species which may conceivably be in any ecologically-significant danger from toxicants associated with crude oil or dispersants are those which form fairly localised aggregations.[6] Arrow-worms, fish eggs and larvae may come into this category. The likelihood of ecologically-significant effects in such cases is examined in detail in Section 2 (Fisheries).

Finally, to keep our assessment of toxic effects on plankton in their true perspective it is helpful to consider the scale of plankton production in the sea relative to that of petroleum spilled. Fraser[10] estimates annual world phytoplankton production to be about 150,000 million tons and that of zooplankton 15,000 million tons. By contrast estimates of annual losses of petroleum in the sea vary from 2.5 million to an improbable maximum of 10 million tons. Only a fraction of this petroleum is toxic to plankton and very little of this is ever likely to come into contact with planktonic organisms.

References

1. NELSON-SMITH, A., *Oil pollutions and marine ecology*, Plenum Press, New York, 1973.
2. OTTWAY, S. M., *The comparative toxicity of crude oils; The ecological effects of oil pollution on literal communities*, Institute of Petroleum, London, 1971, pp. 172–180.
3. BROWN, S. O. and REID, B. L., *Experiments to test the diffusion of oxygen through a surface layer of oil*, Texas A & M Research Foundation, College Station, 1951.

4. SMITH, J. E. (Ed.), *Torrey Canyon pollution and marine life*, Cambridge University Press, 1968.
5. CLARK, R. B., *Oil pollution and its biological consequences, a review of current scientific literature*, Prepared for Great Barrier Reef petroleum drilling Royal Commissions, 1971, 111 pp..
6. MOORE, S. F., CHIRLIN, G. R., PUCCIA, C. J. and SCHRADER, B. P., *Potential biological effects of hypothetical oil discharges in the Atlantic Coast and Gulf of Alaska*, Report prepared by the Massachusetts Institute of Technology for the National Oceanic and Atmospheric Administration, 1974.
7. WILSON, K. W., 'The toxicity, of oil-spill dispersants to the embryos and larvae of some marine fish,' *FAO Tech. Conf.Mar.Pollut.*, Rome, 1970, Paper E-45.
8. GABRIEL, P. L., DIAS, N. S. and NELSON-SMITH, A., 'Temporal Changes in the Plankton of an Industrialised Estuary', *Estuarine and Coastal Marine Science* 3, 145–151 (1975).
9. SANBOURN, HERBERT R., 'The effects of petroleum on ecosystems', in *The Effect of Petroleum on Arctic and Sub-Arctic Marine Environments and Organisms*, (Donald C. Malins, Ed.) Academic Press, 1977.
10. FRASER, J. H., *Nature Adrift, the Story of Marine Plankton*, G & T Foulis, London, 1962.

2. Fisheries

Theoretical background

As was said earlier, the purpose of this chapter is to arrive at an assessment of the effects of 'oil pollution', an environmental stress, for which man is largely responsible, on the biology of the sea. It is important, in this context, to consider the particular case of sea fisheries, a sector of the marine environment in which man already exerts a direct and measurable influence. The term, 'fisheries', has a variety of meanings, the widest embracing all exploited stocks of marine organisms from macro-algae to whales, and the narrowest exploited species of fish alone. In this chapter, the term is taken to mean the fish and shellfish of commerce.

Oil pollution is a potential threat to these resources for three main reasons:

(i) Pollution may directly harm fish and shellfish stocks.
(ii) Pollutants adhering to, or accumulated by, fish and shellfish may render them unacceptable to the consumer.
(iii) The presence of spilled oil may interfere physically with fishing operations.

If a 'stock', by which term fishery biologists usually mean the total biomass of that part of a population which is vulnerable to fishing, is reasonably stable over a representative series of seasons, it follows that the forces tending to increase it are in balance with those tending to decrease it. For an unfished stock, or one which has not been fished for many years, this means that, on average, the gains in weight brought about by the summed growth of the individual fish in the stock, together with the gains in weight attributable to the annual recruitment of young fish, balance the losses in weight caused by natural mortality. If to natural mortality is added mortality due to a given rate of fishing, the immediate effect is to reduce the weight of the stock. Contrary to lay expectation, this reduction is rarely catastrophic, but proceeds until the stock re-stabilises at a new, lower level. In other words, a stable fish population, whether it is unfished or is subjected to a constant, rational level of exploitation, is in a state of dynamic equilibrium. Using the simple notation of Russell[1] the unfished situation may be represented by the expression,

$$M = A + G$$

in which M represents losses due to natural mortality, A gains from the recruitment of young fish to the stock and G gains from the growth of the fish in the stock. Similarly, the stable, fished situation may be represented by the expression,

$$C + M = A + G$$

in which the new term C represents losses to the stock caused by fishing, i.e. the catch. Both expressions represent stable, fish populations but, as we have seen, in the latter case, the standing stock is smaller. One of the tasks of the fishery biologist is to define the rate and pattern of fishing which gives the maximum sustainable yield from the fish population. Under these conditions, compensatory changes in the rate of growth of the stock are at their peak, so maximising the production of fish flesh per unit time. For most commercially-exploited species, an increase in production due to the weeding out of old, biologically inefficient fish, so leaving more food for the faster-growing and more efficient younger fish, is the most important compensatory change. Recruitment tends to vary, often widely, about a mean, not necessarily related to the size of the adult stock. This is because density-dependent mortality, which is strongest during the larval phase, tends to compensate for changes in the numbers of eggs produced and thus for changes in the size of the adult breeding stock.[2–5] In certain cases, the effect of fishing on an unfished stock may even lead to increased recruitment. Under conditions of biological overfishing, a state easiest to attain with fish of low fecundity, the cushioning effect of density-dependent mortality is no longer sufficient to prevent a fall in recruitment.

During the larval phase, natural mortality is high and may be as much as 80% per month.[6] That the loss of many millions of larvae does not lead to extinction is due, firstly, to the fact that fish and shellfish in which such larval losses are normal are highly-fecund and, secondly, because this initial high mortality has decreased markedly by the time the young fish are in balance with their food supply on the nursery grounds, by which time, year-class size has been determined.[5] An idea of the fecundity of some common species of fish is provided by the following data.[7]

Species	*Number of eggs*
Ling	12 – 28 million
Cod	1.8 – 9 million
Saithe	2 – 8 million
Haddock	12,000 – 3 million
Tusk	0.75 – 2.25 million
Turbot	8 million
Flounder	0.56 – 1.6 million
Plaice	16,000 – 346,000
Capelin	15,000 – 53,000
Sardine	15,000 – 55,000
Sandeel	13,350 – 36,300
Herring	12,000 – 33,000

Oil pollution effects on fish and shellfish populations

Having briefly and superficially reviewed the theory of fishing, we are in a position to consider the effects of petroleum, dispersants, etc. on fish and shellfish populations. It will be clear that toxic effects on the eggs and larvae of fish and shellfish which are not in a stock/recruitment relationship are likely to affect recruitment only if the mortality is massive in scale. This would demand, firstly, that the oil spill was very large and that a substantial proportion of the toxic components of the oil had not evaporated but were dissolved in the water and, secondly, that the eggs and larvae were concentrated in the area of the spill. The likelihood of this combination of circumstances being realised is small and it is primarily for these reasons that fisheries' scientists are in general agreement[8–10] that oil floating on the open sea is rarely a hazard to fish and shellfish stocks. Even for those few stocks which are in a stock/recruitment relationship or suffer from restricted spawning grounds, the chances of spilled petroleum offering a threat to recruitment is small. Indeed, there is no evidence anywhere in the literature, that floating oil has ever affected the recruitment to any marine fish or shellfish stock. One of the few cases when pollution-induced mortality of fish eggs and larvae was recorded at sea followed the massive use of high toxicity dispersants during the *Torrey Canyon* incident but even then, the effects were so local that no effect on the subsequent recruitment of fish or crustacea has since been detected.

If it is accepted that oil and dispersant pollution of the open sea is rarely an ecologically significant threat to fish, shellfish eggs and larvae, does the same also apply to later stages in their life history—stages from the O-group onwards—where, in all probability, year-class size has been determined? In theory, a further increment to mortality at this stage would resemble the effects of fishing and, moreover, fishing without reference to mesh size regulations. Russell's equation could thus be rewritten to include the population mortality term, P:

$$P + C + M = A + G$$

How significant a threat this term P implies would depend on the toxicity of the pollutant, the amount released and the degree of concentration of the threatened resource. The greatest dangers probably lie in petroleum spilled directly into shallow, inshore waters. The latter include important nursery grounds for flatfish such as plaice and turbot. Since year-class size is probably fixed quite soon after the arrival of young fish on the nursery grounds, any increment to natural mortality would have an effect on recruitment to the fishable stock. The extent of the effect would depend on the extent to which toxic components were dispersed in the water, the size of the area affected relative to the total nursery area of the species and the size of the spill. This is a theoretical possibility which cannot be ignored; it is sufficient reason to ban the use of high toxicity dispersants in shallow, nursery ground areas.

The toxicity of petroleum and dispersants to adult fish and shellfish has been well-reviewed.[1] Adult fish in the open sea are unlikely to be sufficiently concentrated for toxic effects to be of ecological significance. There is no evidence that even such major oil spills as the *Torrey Canyon* disaster[11] or the Santa Barbara blowout[12] have killed adult fish in the *open* sea. One of the few cases where spilled oil was shown to kill large numbers of fish was at Buzzards Bay, Massachusetts in 1968.[13] In this case, the oil was a highly toxic light fuel spilt directly into turbulent shallow water. The *Amoco Cadiz* oil spill in March 1978 is also of interest in this respect. This ship was wrecked on the Brittany coast and lost its entire cargo of 100,000 tons of light Arabian and 123,000 tons of light Iranian crude oil over a period of some two weeks. Rough seas dispersed the oil in the sub-littoral zone and dead fish were seen for some 10 km around the wreck.[14] Virtually all of the fish killed belonged to inshore species.

There are two groups of fish which pass, both as adults and young, through the confined waters of estuaries. They are called catadromous fish, for example eels, which spawn in the sea and feed mainly in freshwater, and anadromous fish, for example, salmon and sea trout, which spawn in freshwater but feed mainly in the sea. The concentration which this implies makes such species especially vulnerable to the effects of pollution, as many former salmon runs lost from industrially-polluted rivers testify. There is little evidence that estuarine pollution by petroleum has damaged such fisheries but it is a theoretical possibility, especially for species like some Pacific salmon and smelt which often spawn in estuaries. Similar considerations apply to sublittoral shellfisheries where, to the danger of mortality, may be added that of tainting. The latter is a problem with fishing gear as well as with the stock itself. Thus Spooner[15] noted that fixed net contamination was the main fisheries effect of a moderate spill in Tarut Bay, Saudi Arabia. The use of sinking agents to prevent oil coming ashore has also been shown to contaminate fishing gear[16] and to lead to mortality and tainting of benthic animals and thus the food of demersal fish. J. de Veen (personal communication) found that following the sinking, by amine-treated sand, of 100 tons of petroleum mixture comparable with partly-weathered crude oil, the abundance of burrowing sea urchins, *Echinocardium cordatum*, was reduced and there was evidence that this species and the common starfish, *Asterias rubens*, had eaten the oil. Edible crabs, *Cancer pagurus*, and other crustacea, such as the lobster, *Homarus gammarus*, have been observed to eat sunken oil in laboratory aquaria. In practical terms, the problem of tainting is probably the most frequent, tangible effect of spilt oil on the fishing industry.[9,11] It is a problem, not because tainted fish and shellfish are a significant threat to public health as their unacceptable taste would probably be sufficient in itself to prevent ingestion,[17] but because tainting and fear of tainting can have serious effects on marketing.

A final, long-term perspective on the problem of oil pollution and fisheries may be obtained by examining the fisheries statistics of areas, such as the Gulf of Mexico, Lake Maracaibo, the North Sea and Baltic Sea, which have been subject to oil pollution for many years. Caution must be exercised in this type of approach, because landings may not relate in the short term to abundance and all fisheries statistics are no better than the system used to collect them. Nevertheless, if the offshore petroleum industry were affecting fish and shellfish stocks in a major way, then it might be expected that some indication of this would show up in catch statistics. According to Gusey and Maturgo,[18] total catches in the Gulf of Mexico have increased steadily since 1939, despite the presence of 11,500 oil wells. A big game fishing centre based on species attracted to oil platforms (tuna, marlin (2 species) swordfish and sailfish) has been established 100 miles off New Orleans. Lake Maracaibo in Venezuela is reputed to be one of the most heavily oil-polluted areas in the world. Although described as a 'lake' Maracaibo is connected to the sea by a narrow entrance at Bahia del Tablazo. Its waters are saline with a gradient running from north to south. According to Nemoto[19] there has been no decline in the fisheries, the total catch from which totalled 24,000 tons in 1970.

The North Sea and Baltic Sea have been polluted by refinery discharges, intentional discharges from ships and many accidental spillages from the beginning of this century. Both are comparatively shallow seas supporting substantial fisheries which might be expected to decline as a result of oil and other forms of pollution. In fact, the contrary is the case; total landings have shown an increasing trend over the last five decades.[20] Although during that period, there have been major fishing and environmentally-induced fluctuations in the fish stocks, particularly in the relative abundance of pelagic and demersal species, and in the abundance of all species during and after the two world wars, there is no evidence for an oil pollution-induced decline in the productivity of these seas.

References

1. RUSSELL, E. S., 'Some theoretical considerations on the "overfishing" problem', *J.Cons.perm. int. Explor. Mer.* **6**. 3–20 (1931).
2. RICKER, W. E., 'Stock and recruitment', *J. Fish. Res. Bed. Can.* **11**, 559–623 (1954).
3. BEVERTON, R. J. H. and HOLT, S. J., 'On the dynamics of exploited fish populations', *Fishery Invest., London. Ser. 2* **19**, 533 (1967).
4. CUSHING, D. H. and BRIDGER, J. P., 'The stock of herring in the North and changes due to fishing', *Fishery Invest., London, Ser. 2* **25**(1), 123 (1966).
5. CUSHING, D. H., 'The natural regulation of fish populations', in *Sea Fisheries Research*, (F. R. Harden Jones, Ed.), Paul Elek (Scientific Books), London, 1974, pp. 399–412.

6. HARDING, D., 'The biology of fish larvae with particular reference to the plaice (*Pleuronectes platessa Linn.*); in *Sea Fisheries Research*, (F. R. Harden Jones, Ed.), Paul Elek (Scientific Books), London, 1974, pp. 55–66.
7. SIMPSON, A. C., 'The pelagic phase', in *Sea Fisheries, their investigation in the United Kingdom*, (M. Graham, Ed.) Edward Arnold (Publishers), 1956, pp. 207–250.
8. KORRINGA, P., 'Biological consequences of marine pollution with special reference to the North Sea fisheries', *Helgolander wiss. Meeresunters* **17**, 126–140 (1968).
9. SIMPSON, A. C., 'Oil emulsifiers and commercial shellfish', *F.d. Studies* **2** (Suppl.), 91–98 (1968).
10. COLE, H. A., *Marine Pollution*, Oceanology International, Conference technical sessions, day 4 session H, 1969.
11. SIMPSON, A. C., *The Torrey Canyon disaster and fisheries*, Ministry of Agriculture, Fisheries and Food (Lab Leaflet No. 18), London, 1968.
12. STRAUGHAN, D., *Oil pollution and fisheries in the Santa Barbara Channel: Biological and oceanographical survey of Santa Barbara Channel oil spill 1969–70*, Vol. 1, *Biology and bacteriology*, Allan Hancock Foundation, University of Southern California, 1971, pp. 245–254.
13. HAMPSON, G. R. and SANDERS, H. L., 'Local oil spill', *Oceanus* **15**, 8–11 (1969).
14. SPOONER, M. F., '*Amoco Cadiz* oil spill', *Mar.Poll. Bull.* **9**, No. 11, 281–311 (1978).
15. SPOONER, M. F., 'Oil spill in Tarut Bay, Saudi Arabia, *Mar. Poll. Bull.* **1**, 166–167 (1970).
16. WARDLEY-SMITH, J. and SHELTON, R. G. J., 'Disposal of oil spills from water surfaces', *Proc. 8th World Petroleum Congress*, 1971, pp. 67–76.
17. MARTIN, A. E., 'Water pollution by oil—some health considerations', in, *Water Pollution by Oil*, (P. Hepple, Ed.), The Institute of Petroleum, London and Elsevier, Amsterdam, London, New York, 1971.
18. GUSEY, W. F. and MATURGO, Z. D., *Petroleum production and fish and wildlife resources—The Gulf of Mexico, Louisiana.* Environmental Conservation Dept., Shell Oil Co., New York, 1971.
19. NEMOTO, T., *La Pesca en el Lago de Maracaibo.* Informe Technico No. 24 Caracas, 1971.
20. COLE, H. A. and HOLDEN, M. J., 'History of the North Sea Fisheries 1950–1969', in *North Sea Science*, (E. D. Goldberg, Ed.), M.I.T. Press, Cambridge, Massachusetts and London, England, 1971, pp. 337–360.
21. NELSON-SMITH, A., *Oil pollution and marine ecology*, Plenum Press, New York, 1973.

3. The shore and splash zones

The shore is the most obvious area to suffer from oil pollution. Spilt crude, or products such as fuel oils, may be stranded, covering rocks, sand, plants and animals with an obvious dark film. Weathered lumps of tar, often resulting from distant spills or chronic discharges, drift in and are stranded. Oily effluents may be discharged on or near the shore. All these things can be seen, smelt or felt by the general public, especially when on holiday, so from an amenity point of view it is right that beaches should be cleaned. However, cleaning procedures probably cause as much concern among biologists as the oily pollutants themselves and, in some cases, cleaning increases damage to shore life. Thus there may be conflict between amenity and biological interests or between one kind of amenity interest and another. For example, flats of fine sand might look much better and be more pleasant to sit on if oil films were chemically cleaned or if the surface was removed using bulldozers, but bait-diggers or cockle fishers are likely to find that such treatments reduce the numbers of worms and shellfish more than a surface oil film.

Such conflicts are best resolved by shore classification schemes. In Milford Haven, for example (see Fig. 4.2), popular tourist beaches are cleaned, the rocky shores that are seldom visited are not cleaned and remaining areas may or may not be treated following consultation with appropriate organisations such as the sea fisheries committee and the local field centres.

The correct control and treatment of shore oil pollution involves the asking of many questions. What are the short- and long-term effects of a particular type of oil and how long will it remain on the shore if not treated? How efficient are different cleaning methods and what damage do they do to shore life? Where is the best place to discharge ballast water or effluent from a refinery, and what is the effluent quality that is acceptable for a particular area? These and related problems are being investigated by research workers in a number of disciplines and their results should, and in some cases do, form the basis for environmental decision-making. There are four main approaches to these problems.

First, surveys may be carried out following oil spills. Large spills such as the *Amoco Cadiz*[1] have been particularly well documented but there are several other studies of importance because they cover a range of oils, treatments and biological communities[2–4] or have been carried out over a long period of time. One of the earliest post-spill surveys was that following the *Torrey Canyon* in 1967,[5] so it is of particular interest that some effects were still

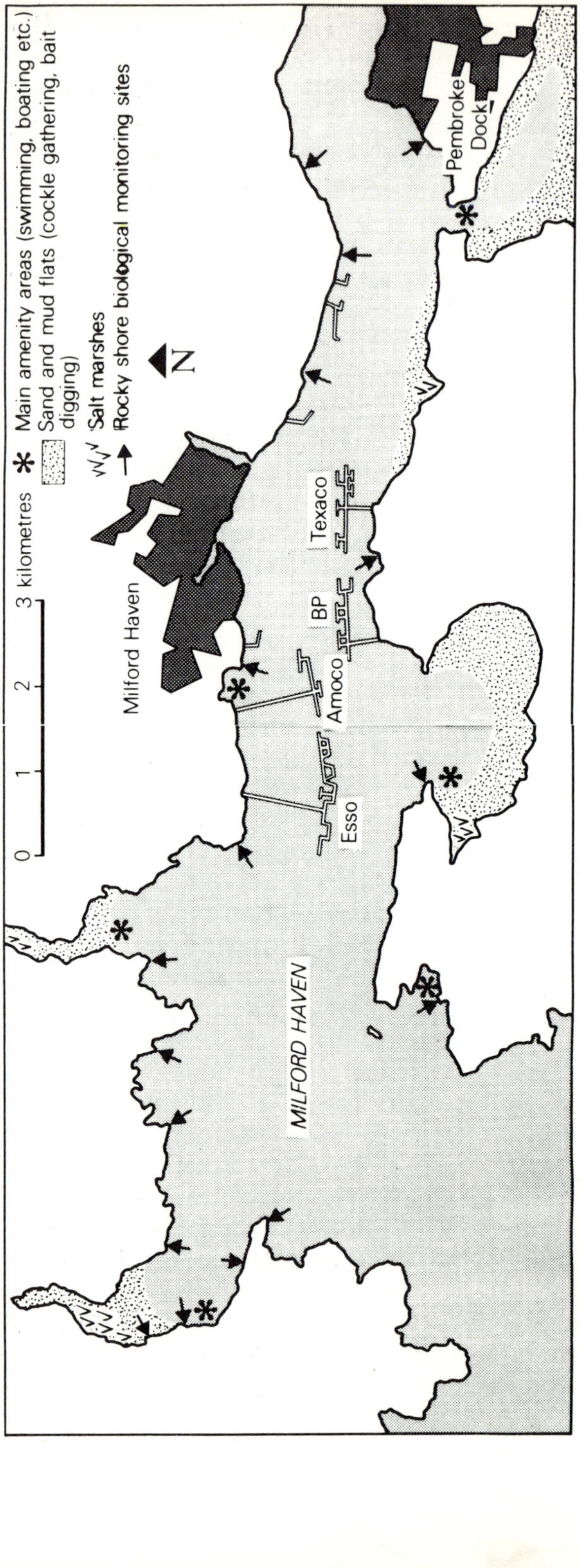

Fig. 4.2 Map of Milford Haven

detectable in 1977.[6] A comprehensive summary of spills up to 1975 is given by Ottway and Knight,[7] and in recent years a useful and continuing source of information is the series of biennial API/EPA/USCG Oil Spill Conference papers.[8] Most post-spill surveys suffer from the problem of inadequate pre-spill biological data—if marine biologists could predict where major spills were going to occur they would probably be in other jobs!

Second, long-term monitoring schemes are being increasingly used in new oil development areas such as Sullom Voe.[9] They are useful for finding the long-term effects of usually small oil spills, cleaning treatments, refinery effluents and ballast water discharges, all of which are likely to occur in estuaries used as oil ports.

Third, field experiments involving oiling and/or cleaning treatments may be carried out on different types of shore.[10–13]

Fourth, laboratory tests are used for a number of different purposes:

(i) Comparing the toxicity of different oils and dispersants. Many different species have been used in such toxicity ranking tests and criteria for toxicity are usually, various behavioural responses (e.g. limpet 'drop-off', retraction of winkles into their shells) or death.

(ii) Comparing the relative susceptibilities or tolerances of a range of species to a particular pollutant. For example, the effects of a dispersant were studied on a range of intertidal animals,[11] and on a range of saltmarsh plants.[10] These studies showed that some species are much more resistant than others.

(iii) Studying the effects of oils and dispersants on the behaviour, growth or metabolism of various species.

(iv) Studying the uptake, accumulation, retention, food-chain transfer, breakdown or excretion of hydrocarbons by various species.

The use of different types of test is discussed by Davis,[14] and toxicity testing of oil spill dispersants for licensing purposes by Norton and Franklin.[15]

All these various laboratory and field approaches are of importance when making ecological predictions, for it can be misleading to rely on one type of information, for example, toxicity ranking data only.

Information obtained using these various methods shows that the range of effects of oil and dispersant pollution on shore organisms is very large. On the one hand some types of chronic oil pollution or misuse of some dispersants can lead to the virtual elimination of most species for varying lengths of time. On the other hand, some cases of oil pollution have produced very little damage on the shore and the growth of saltmarsh grasses may actually be improved, lush green growth appearing after oiling with some crudes and residues.

There is considerable published material concerning the effects of oils and dispersants on marine plants and animals, and no attempt will be made here to present a comprehensive review as several are already available, (for example, see Refs 16–18). Instead, the factors which appear to be important in determining the extent of damage will be discussed. Some of these factors are related to the oil and cleaning treatment while others are related to biological and other natural circumstances. The former include volume and type of oil, and cleaning treatments, the latter include the topography, hydrography and climate of the affected area, the nature of the biological communities there, and the season of the year.

Volume of oil

It is hardly necessary to say that the greater the volume of oil discharged the greater the potential damage to the shore. There are some points, however, which are not so obvious. If the oil is fairly light, for example a typical Middle Eastern crude oil, there is a limit to the amount which will 'stick' to the shore, and excess will run off and be carried elsewhere, perhaps to more distant shores, by the tide and the wind. Damage caused by the oil remaining on any particular shore is therefore more likely to be related to the toxicity of the oil and the biological community. If the oil is thick and viscous, for example heavily-weathered crude oil, heavy residual fuel oils, or mousses, then much larger amounts can stay on a shore, and plants and animals may be smothered under a sticky blanket.

Type of oil

A variety of crude oils and oil products have now been studied in various ways, and it is evident that both crude oils and products differ widely in toxicity. Toxicity does not vary consistently with oil properties but, in general, it appears that severe, toxic effects are associated with low-boiling compounds and aromatics. With plants (which have been studied with herbicidal oils in mind) there is experimental evidence that toxicity increases along the series alkanes (paraffins) – cycloalkanes (naphthenes) – alkenes (olefins) – aromatics.[19,20] Within each series of hydrocarbons the smaller molecules are more toxic than the larger.

These results are in general agreement with experiments on animals (usefully summarised by Craddock[21]). Observations following oil spills suggest that the greatest short-term damage has been caused by lighter oil, particularly when confined in a small area, for example the diesel fuel from the *Tampico Maru,*[2] the No. 2 fuel oil from the *Florida*,[22] and the gasoline from the *Dona Marika.*[4] Severe immediate effects (in the case of the *Dona Marika*,[4] reduction of the limpet population over about one mile of shore from over 50 per dm^2 to 1–10 per dkm^2) do not appear to preclude good, if not complete, recovery over periods of time ranging from two to ten years.

Spills of heavy oils, such as Bunker 'C' Fuel Oil,[3,46] may blanket areas of shore and kill organisms mainly through smothering, (which is a physical effect), rather than through acute toxic effects. Crude oils, when fresh, may cause acute toxic damage, but as they weather, the fractions responsible for such damage will decrease and later effects are more likely to be related to the physical properties of the residues. In general, crude oil spillages appear to have resulted in fewer mortalities of inter-tidal organisms per unit area than some product spillages.

Long-lived oils, and oil incorporated into sediments (which sometimes includes more toxic, lighter oils), may have longer term chronic effects, and some concern has been expressed in particular over uptake of hydrocarbons including carcinogenic polycyclic aromatics. There is much evidence in the literature consistent with the generalisation that marine organisms exposed to petroleum hydrocarbons will rapidly accumulate these compounds in the tissues but will gradually release most of the quantities taken up once the source of contamination has been removed. The subject has been reviewed by Varanasi and Malins,[23] and more recently by a Royal Society working group.[18] The latter concluded that there was, so far, no convincing evidence to support the suggestion that polycyclic aromatics from spilled oil could be the direct cause of outbreaks of cancer and other diseases in marine organisms. There is, however, virtually no information on the accumulation, metabolism or effects in marine organisms of products arising from the chemical or microbial oxidation of oil.[24]

Frequency of oiling

Incidents described in the literature indicate that most types of shore recover well, though not necessarily completely, in periods ranging from about two to ten years from one oil spillage incident.

Some species on the shore may survive the oil, this usually being the case with the common brown seaweeds which have a protective, slimy covering. Others, such as limpets and barnacles, may die but are replaced during the shore recovery period by settlement from the young planktonic stages of these animals.

Estuaries used as oil ports suffer from comparatively large numbers of small spillages. Problems arise where many such spillages affect the same area of estuary, because in such a case the shore life may never have a chance to recover properly. Thus many small successive spillages in the same place are likely to be more damaging than one larger spillage. Experimental work[10] has shown that many saltmarsh plants, including the widespread grass *Spartina anglica*, can withstand four successive monthly oilings with fresh Kuwait crude oil, but decline with further oilings. Death can result in bare mud liable to erosion.

The greatest frequency of oiling may be found near continuous discharges such as refinery effluents. Effluents which receive only primary treatment (gravity separation) commonly have oil contents between 10 and 25 parts of oil per million of water (ppm). In calm conditions thin oil films may rise to the water surface in the discharge area, and be left on the shore as the tide goes down. If the effluent is discharged into a saltmarsh area, the marsh plants (to which oil sticks firmly) may die following repeated exposure to oil. One area where this happened is now showing good recolonisation following improvements in effluent treatment.[9,25] The continuing performance of the marsh plants and animals is seen by the company concerned as a useful monitor of effluent quality.[26]

A refinery effluent in a fairly sheltered rocky bay in Milford Haven has had the effect of locally reducing the population of the limpet *Patella vulgata* and therefore increasing the abundance of green and brown seaweeds. Top shells, winkles and barnacles are also reduced in this area. In this particular case, however, it would be wrong to jump to the conclusion that oil alone is to blame—the effluent is made up mainly of fresh water, and marine animals do not like that very much either!

State of oil

A given volume of an oil may have different effects according to whether it reaches living organisms as thin oil films, thick oil films, water-in-oil emulsions or oil-in-water emulsions. Oil films or water-in-oil emulsions (i.e. mousses) may coat organisms on the shore and have a directly toxic effect or a physically smothering effect. Oil-in-water emulsions may be formed from oil slicks, naturally or following the use of dispersants, and may be present in a variety of industrial and other effluents from land-based discharges. Such emulsions, which may disperse through large volumes of water, are more readily available to bacteria for degradation, but oil in this state is also more likely to be ingested by plankton (including planktonic larval stages of some shore organisms) and a variety of animals which feed by filtering particles from the water. Shore animals of this type include barnacles and mussels.

Interactions affecting oil toxicity

The toxicity of oil to shore organisms may be altered through interactions with other compounds and this is of particular importance when considering effluents which contain low concentrations of oil and other compounds. Factors which may cause variations in toxicity of the oil by increasing or reducing the stress to particular animals include salinity, pH, sulphides, phenols, ammonium compounds, suspended material, dispersants and temperature.[27]

The large number of hydrocarbons and other chemical compounds contained in oils and in refinery effluents yields a great range of possible interactions which so far are poorly understood. Furthermore, it is not known whether ecological changes, where present, result from above-specification peaks of effluent compounds, or from continual low-level discharge. The formulation of some general principles applicable to the discharge of mixtures of oil and other substances into receiving water would be preferable to an infinity of *ad hoc* tests of individual effluents.

Cleaning treatment

In practice, two approaches are commonly used for cleaning oiled shores, namely physical removal (including cutting of oily vegetation) and dispersal.

Burning is an initially attractive course of action, but anyone who has thrown a lighted match on to stranded oils will conclude that they do not burn easily, if at all. This is because of evaporative losses and absorption of water. The use of petrol or flame-throwers to encourage the process is clearly not very good for life on the shore.

Physical removal of oil and oily substrates, either on a large scale using bulldozers or on a small scale using buckets and spades, is commonly used on amenity sand and shingle areas.

Such areas are not usually the richest of shores biologically, and physical removal appears to be acceptable from both the biological and amenity points of view. This method is not usually suitable, however, on rugged rocky shores or soft mud or saltmarsh substrates. Appropriate machinery cannot easily be manoeuvered along these types of shore, and removal of surface layers of sediment from mud flats and marshes would probably mean removal of many animals such as worms, and root systems of plants. This method is only likely to be justifiable in cases of thick smothering by potentially long-lived deposits which would prevent re-colonisation by plants and animals. In the case of lightly oiled saltmarshes, the cutting and removal of oily stems and leaves (not the wholesale removal of the marsh surface) seems to be an acceptable treatment provided that heavy trampling damage can be avoided.

Misuse of dispersants can give rise to many problems and the severe and sometimes long-lasting effects of products used to clear the *Torrey Canyon* oil are often quoted. Some later products have been demonstrated to be substantially less toxic,[13,28] and if used in the correct way on rocky shores should not significantly increase damage already done by the oil. Whatever the toxicity of a dispersant, however, its use on mud and saltmarsh areas is likely to cause problems. This is because dispersed oil may sink more readily into sediments, where it may remain for a long time with unknown effects.

Misuse of dispersants during transport and handling can sometimes affect areas above high tide mark. Spillage of comparatively toxic products on cliffs during the *Torrey Canyon* cleaning operations damaged several areas of cliff vegetation, and recolonisation was slow.[29] Spraying the shore or near-shore waters, indiscriminately or in windy conditions, can result in dispersant reaching lichens and other plants growing on cliffs, with little likelihood of dilution with adequate quantities of seawater.

Topography, hydrography and climate

With some spillages, for example, the *Tampico Maru* and the *Dona Marika*, oils have been spilt and more or less confined in bays through various circumstances including prevailing wind direction. Some refinery effluents are discharged into small bays or creeks where the hydrography does not favour rapid dispersal.[26,27] These various conditions all help to produce localised effects of varying severity. Conversely, good dispersal minimises effects on the shore (although of course it may add to offshore effects). The oil port of Milford Haven is an inlet with good tidal flushing, and oil spills, dispersant use and refinery effluent discharges since 1960 have not caused an overall decline of shore organisms. Localised effects have occurred in places and of these the accumulation of oil at a small saltmarsh and the effects of a refinery effluent in a rocky bay are likely to persist for some time. The areas affected are small relative to the length of coastline within Milford Haven. It should be pointed out that results for the Haven cannot necessarily be extrapolated for making predictions about other inlets with possibly different hydrographic and biological conditions.

Climatic and weather conditions can influence the course of events following a spill by determining evaporation rates (wind and temperature), formation of emulsions (wind and waves), viscosity of stranded oil (temperature) and degradation rate of oil (temperature and sunlight). The influence of these factors on the fate of oil is discussed in Chapter 2.

Strong winds can blow oil above the high tide mark where it may adhere to cliff plants, lichens or sand dune plants. Mousse from the *Metula* spill in the eastern Straits of Magellan in 1974 was in places blown ten metres into sand dunes dominated by the grass *Leymus*. In this and similar cases re-growth has proved possible, provided the oil has not penetrated into the root systems and does not form a thick smothering deposit.[30]

As well as pollutant, cleaning and geographical factors, the extent of damage depends on a number of biological factors, the most important of which are concerned with the type of community and the wide differences in the possible susceptibility of different species. As examples of this, a number of different communities will be described.

Rocky shores

With an exposed limpet/barnacle-dominated shore (Fig. 4.3) such as is common in north west Europe, some oils and/or dispersants can cause substantial limpet 'drop-off'.[31] Although many detached limpets are capable of recovery in the laboratory, they do not get a chance in real life because of predation by gulls. This was particularly striking after the *Dona Marika* accident, when the affected area was well defined by the flocks of gulls circling the bay and picking up limpets as they dropped off the rocks. Limpets graze seaweeds (see Fig. 4.4) which are relatively resistant to oil and dispersant pollution. When the limpets have gone, therefore, (see Fig. 4.5) there is an invasion of the stringy green weed *Enteromorpha* followed by the brown weed *Fucus* (see Fig. 4.6). The extensive weed growth in turn tends to smother and prevent the settlement of barnacles and other species.

Fig. 4.3 An exposed limpet/barnacle-dominated rocky shore

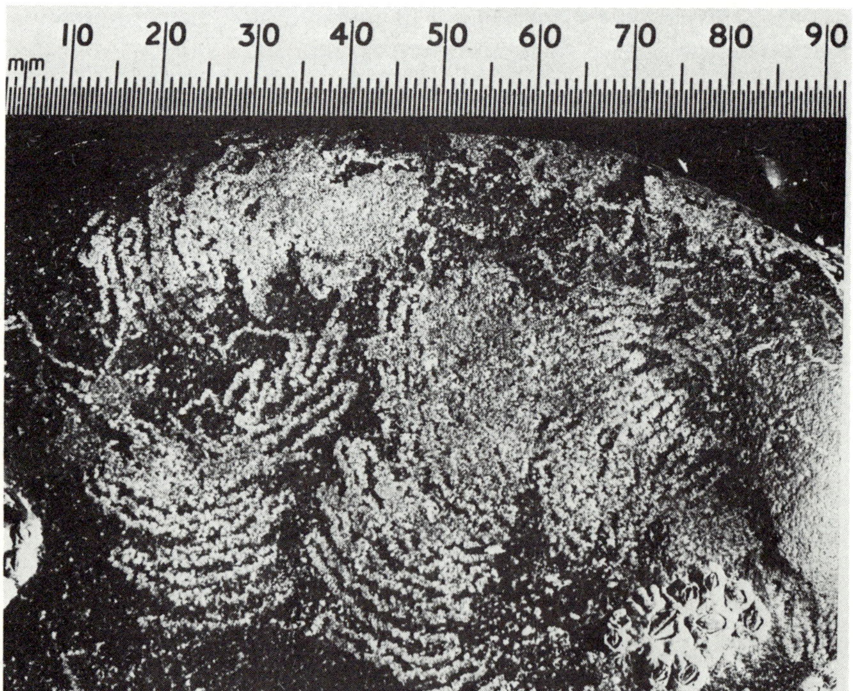

Fig. 4.4 Grazing tracks of limpets

Fig. 4.5 Scars left by limpets which have dropped off

Fig. 4.6 Rock covered with green weed *Enteromorpha* (right) and brown weed Fucus (left)

Gastropod molluscs such as winkles (*Littorina* spp.), top shells (*Gibbula* spp.) and dog whelks (*Nucella* spp.) commonly retract into their shells under the influence of oils and dispersants and are then liable to be washed away by tidal action. Re-colonisation of shores by adults has been observed.[32] Crapp deduced that animals were washed into the sub-littoral zone and eventually crawled back into the littoral.[11]

A sheltered temperate rocky shore naturally has few limpets and is often dominated by large masses of weed such as the egg or knotted wrack, *Ascophyllum*, which (naturally) cause reduced barnacle populations. Thus, some characteristics of a polluted exposed rocky shore are close to the characteristics of a clean sheltered rocky shore. Pollution of such a clean sheltered shore is therefore not likely to produce such large-scale effects as comparable pollution on an exposed shore.

Unlike many temperate shores, the upper part of the intertidal zone of tropical rocky shores is often dominated by oysters. Barnacles, winkles, limpets, coralline algae and the brown alga *Sargassum* also occur.[33,34] Very little is known about the effects of oil pollution on such shores.

Saltmarshes

Saltmarshes occur up to the extreme high water mark on sheltered shores and in estuaries in the temperate regions of the world. They tend to act as oil 'traps' because of the sheltered conditions and the natural absorptive capacity of saltmarsh vegetation.

Saltmarshes are often important feeding grounds for wildfowl and waders, and are a source of the organic detritus upon which many estuarine and near-shore food chains are based. Saltmarshes are also of interest in terms of shellfish culture, waste water assimilation and erosion control along estuarine shorelines.

Small succulent glassworts (*Salicornia* spp.), are common lower-level annual saltmarsh plants, frequently being the colonisers of stretches of mud flat. An oil film covering these plants kills the shoots, and being annuals, they have no large underground systems from which new shoots can arise. An oil spill on such a community therefore kills a high percentage of the plants.

Another common lower saltmarsh plant in Europe is the grass *Spartina anglica*, a perennial with a large underground system of roots and rhizomes. A similar species, (*Spartina alterniflora*) occurs in North America. An oil spill usually kills the shoots, but new growth usually occurs from buds protected just below the soil surface. The plant is not, however, resistant to chronic pollution, perhaps because oil interferes with the oxygen diffusion pathway from the shoots down to the roots in their anaerobic environment. As *Spartina* does not commonly associate with other flowering plants, its death results in exposure of bare mud.

The upper saltmarsh is often relatively diverse and in parts of North West Europe, for example, may have a community of the sea rush *Juncus maritimus*, with the grasses *Festuca rubra* and

Agrostis stolonifera, and a number of other plants. One spill is often enough to kill most of the *Juncus* in which case there is an increase of *Agrostis*. This grass is well adapted to the colonisation of dead *Juncus* patches as it has fast-growing creeping stems. Chronic pollution of this sort of community eliminates species according to their tolerance ranking, but some are resistant and can survive at least twelve successive monthly oilings. In contrast to the simple *Salicornia* and *Spartina* communities, the more complex or diverse upper marsh community thus has greater ability to withstand total destruction.

Mangroves

Mangroves are trees or bushes growing up to the extreme high water mark on sheltered shores and in estuaries throughout the tropics reaching their greatest luxuriance in parts of south east Asia. In both hemispheres, there is a zone in which mangroves and saltmarshes can be found growing together—such areas include the Gulf Coast of the USA, southern Australia, and southern Japan. Mangroves are of direct commercial importance in the production of timber, firewood, charcoal and bark for tanning. They are the feeding and breeding grounds for a variety of fish, crustaceans and molluscs, and the trunks and prop roots usually support a varied fauna of oysters, snails, barnacles, crabs and other invertebrates.

Mangrove leaves are an important source of the detritus upon which many marine food chains are based. Mangroves, by trapping and stabilising sediment, increase land area over the centuries and also protect nearshore corals from being overwhelmed with sediment. Some idea of the accretion on mangrove coasts is given by the fact that when Palembang in south eastern Sumatra was visited by Marco Polo in 1292 it was still a coastal or rivermouth port; it is now 50 km inland.[35]

Comprehensive accounts of mangroves are given by Macnae[35] and Chapman.[36] Some effects of petroleum hydrocarbons on mangrove systems reported in the literature have been summarised by Baker *et al.*[37] Far less is known, however, than is the case for saltmarshes. An overall impression is that following severe oil spills, the acute short term effects are likely to be trapping of oil, high mortalities of invertebrates, defoliation of mangroves, and death of seedlings. In the longer term, oil is likely to weather comparatively quickly,[38] and both recovery of damaged mangrove trees and re-colonisation by invertebrates have been observed. In areas where large trees have died (e.g. those described by Baker *et al.*[37]) it is not known how long the re-colonisation and recovery process to mature mangrove forest takes.

As is the case with many saltmarsh plants, mangroves have oxygen diffusion pathways supplying the roots in their anaerobic environment. Oxygen enters these pathways through pores on aerial 'breathing roots' or pneumatophores, and it has been suggested that oiling of pneumatophores damages mangroves by interfering with oxygen supplies to the root system.

Corals

Coral reefs occur in tropical waters and are of three main types—fringing reefs, which occur in shallow waters bordering shorelines; barrier reefs, which are larger and occur further offshore; and atolls. Some shallow water corals may be exposed to the air during low water spring tide periods, and these are likely to be particularly susceptible as oil may be stranded on them rather than floating over them. Corals appear to have a number of behavioural mechanisms which help as a protection against oil; the most important of these is mucus secretion which helps to repel or trap oil. Information on the effects of oil on corals has been usefully reviewed and discussed by Ray;[39] but so far there is relatively little. From the oil spill control point of view, the priority should be to keep oil away from the shallow upper reef areas or quiescent lagoon waters.

These examples from a range of shore types show how effects may vary with shore community and illustrate a number of post-pollution reactions. In all communities time of year and even time of day may also help to determine pollution effects because they control life cycles and activity rhythms. Each shore organism has its own yearly cycle of growth and development, and one stage may be more susceptible than another stage; for example, seedlings of plants, if oiled, are more likely to die than older plants. It has also been demonstrated that many organisms have diurnal and tidal rhythms of behaviour and activity and may be more susceptible at particular times of day. A good example concerns the effects of crude oil on limpet detachment rate. This is greater during the limpets' active phase, when they are crawling around and are more likely to become oiled under their feet. Milford Haven limpets seem most active between midnight and three in the morning, which unfortunately means that biologists using them for toxicity tests also have to be active at this time.

Monitoring

Monitoring, both chemical and biological, is receiving increasing attention in oil ports and industrial development areas. Data on the long term effects of spills, cleaning methods and effluents should be fed back to organisations such as harbour authorities, water authorities and oil companies. Modifications in operations or changes in effluent contents can then be discussed and initiated if initially unpredicted and undesirable effects come to light.

A particularly fruitful approach has been quantitative biological sampling from localised grids or transects of sites around chronic discharges or jetties. In combination with physical and chemical measurements, for example, sediment particle size and hydrocarbon content, this approach allows comparison with changes occurring at reference sites at the edges of grids, the identification of gradients of effect, and correlation between biological, chemical and physical data. Ecological monitoring which is not specific to an effluent discharge or other point source of disturbance presents greater problems in selection of reference or 'control' sites and interpretation of change. Sedentary organisms such as many of the intertidal and sea-bed invertebrates provide the most useful indicators of pollution-related changes because, being immobile, they are unable to escape acute or chronic pollution. Fish or plankton are more difficult to sample quantitatively because of natural daily, seasonal and annual movements. Detailed information on monitoring methods and case history examples is available elsewhere,[26,40] so will not be discussed further here.

Larger animals on the shore

The animals mentioned so far in this section are the small invertebrates which dominate the intertidal zone. The shore or nearshore waters may also be used by a variety of larger animals which are at risk following oil spills.

Seals have been oiled on a number of occasions, for example, elephant seals and California sea lions following the *Santa Barbara* incident,[41–43] and grey seals on Skomer Island (south west Wales) following an oil spill of undetermined origin in 1974[44] and the *Christos Bitas* accident in 1978.[45] Effects on health and survival of adult seals appear to have been small but in the case of the Skomer seals some oiled pups died. In all these cases, however, it has been difficult to separate oiling effects from natural mortalities. A controlled experiment in Canada[43] showed that up to 75 ml of ingested crude oil was not irreversibly harmful to ringed seals but that surface contact had a far greater effect particularly on eyes.

Eleven otters are known to have been killed by the Bunker 'C' fuel oil spilt from the *Esso Bernicia* in Sullom Voe, and a further 18 reported oiled but alive.[46] Attention has also been drawn to the vulnerability of sea otters in Canada.[47] Oil pollution has not had any obvious direct effect on otter, raccoon, skunk, civet, mink, muskrat and nutria of the Louisiana marshes. Increased human activity and possible long term changes in feeding areas are likely to be at least as important in explaining fluctuations in numbers.[48]

In various places throughout the world, domestic animals (sheep, cattle and horses) graze on saltmarshes (see Fig. 4.7). The possibility of contamination of important grazing areas (for example, the marshes of the Burry estuary, South Wales) should be borne in mind. In Shetland and elsewhere sheep sometimes graze seaweeds on the shore, and following the *Esso Bernicia* accident in Sullom Voe, 50 sheep were killed and the fleeces of over 2000 oiled.[46]

Fig. 4.7 Horses and sheep grazing on saltmarsh

Other mammals at risk are whales and polar bears, but there is so far little reliable information concerning oil pollution effects on these species. The same may be said of turtles, *Chelonidae*—aquatic members of the tortoise family, whose limbs are modified into flippers for swimming. One sort, the Hawksbill, *Chelonia imbricata*, yields tortoiseshell while the Green Turtle, *C. mydas*, is edible and used for soup. Indiscriminate harvesting for these purposes has reduced turtle populations disastrously. All species land on sandy shores and scoop out a hole in which they lay their eggs, covering them afterwards with sand. Oil deposited on the shore between the sea and the laying area could coat the adult turtles, possibly to the detriment of their eggs which might get oily while being laid. The young which, after hatching, go down the beach to the sea, might also be oiled and suffer.

References

1. 'Case History—The *Amoco Cadiz*', Oil Spill Conference sponsored by API, EPA and USCG, 19–22 March 1979, Los Angeles, pp. 141–198.
2. NORTH, W. J., NEUSHUL, M. and CLENDENNING, K. A., 'Successive biological changes observed in a marine cove exposed to a large spillage of mineral oil', In *Symposium sur les Pollutions Marines par les Microorganismes et les Produits Petrolièrs*, April 1964, Monaco, 1965, pp. 335–354.
3. CHAN, G. L., (1973) 'A study of the effects of the San Francisco oil spill on marine organisms', *1973 Prevention and control of oil spills conference*, API, EPA, USCG, Washington, 1973, pp. 741–781.
4. BLACKMAN, R. A. A., BAKER, J. M. JELLY, J. and REYNARD, S., 'The *Dona Marika* oil spill', *Mar. Poll. Bull.* **4** (12) 181–183 (1973).
5. SMITH, J. E. (Ed.), *Torrey Canyon pollution and marine life*, Cambridge University Press, 1968, 196 pp.
6. SOUTHWARD, A. J., and SOUTHWARD, E. C., 'Recolonization of rocky shores in Cornwall after use of toxic dispersants to clean up the *Torrey Canyon* spill', *J. Fish. Res. Board Canada* **35**(5), 682–706 (1978).
7. OTTWAY, S. and KNIGHT, M., *Review of world oil spillages. Marine ecology and oil pollution.* Applied Science, London, 1975.
8. API, EPA, USCG, *Proceedings of the 1975, 1977, 1979 and 1981 Oil Spill Conferences.*
9. Field Studies Council Oil Pollution Research Unit, 'Sullom Voe—biological monitoring', in *OPRU Annual Report 1979—1980*, 1980.
10. BAKER, J. M., 'Studies on saltmarsh communities', in *The Ecological effects of oil pollution on littoral communites*, (E. B. Cowell, Ed.) Applied Science, London, 1971, pp. 16–99.
11. CRAPP, G. B., 'Zoological studies on shore communities', in *The Ecological effects of oil pollution of littoral communities*, (E. B. COWELL, Ed.) Applied Science, London, 1971, pp. 102–216.
12. CANEVARI, G. P., 'The restoration of oiled shorelines by the proper use of chemical dispersants', API, EPA, USCG, Oil Spill Conference, 19–22 March, Los Angeles, 1979, pp. 443–446.
13. BAKER, J. M., CROTHERS, J. H., MULLETT, J. A. J. and WILSON, C. M., 'Ecological effects of dispersed and non-dispersed crude oil: a progress report', *Proceedings of the Institute of Petroleum conference on Petroleum Development and the Environment, 20–21 November 1979*. Heyden & Son, London, 1980.

14. DAVIS, J. C., 'Standardisation and protocols of bioassays—their role and significance for monitoring, research and regulatory usage, in *Proceedings of the 3rd Aquatic Toxicity Workshop*, Environmental Protection Service, Halifax, Nova Scotia, Canada. Surveillance Report EPS-5-AR-77-1, 1976, pp. 1–10.
15. NORTON, M. G. and FRANKLIN, F. L., *Research into toxicity evaluation and control criteria of oil dispersants*. Ministry of Agriculture, Fisheries and Food: Fisheries Research Technical Report No. 57, Lowestoft, 1980.
16. NELSON-SMITH, A., *Oil pollution and marine ecology*, Plenum Press, New York, 1980.
17. MALINS, D. C. (Ed.), '*Effects of petroleum on Arctic and Subarctic marine environments and organisms*, Academic Press, New York: San Francisco: London, 1977, Vol. I 321 pp, Vol. II 500 pp.
18. Royal Society of London, 1977, *Oil pollution in the marine environment: some research needs*, A submission to the Royal Commission on Environmental Pollution, 1980.
19. CRAFTS, A. S. and REIBER, H. G., 'Herbicidal properties of oils', *Hilgardia* **18**, 77–156 (1948).
20. HAVIS, J. R., *Herbicidal properties of petroleum hydrocarbons*, Cornell University Agricultural Experimental Station, Memoir, Ithaca, New York, 1950, 20 pp.
21. CRADDOCK, D. R., 'Acute toxic effects of petroleum on Arctic and Subarctic marine organisms', in *Effects of petroleum on Arctic and Subarctic marine environments and organisms*, Vol. II, (D. C. Malins, Ed.). Academic Press, New York; San Francisco, London, 1977, pp. 1–93.
22. SANDERS, H. L., GRASSLE, J. F., HAMPSON, G. R., MORSE, L. S., GARNER-PRICE, S. and JONES, C. C., 'Anatomy of an Oil Spill: long term effects from the grounding of the barge "Florida" off West Falmouth, Massachusetts', *J.Mar.Res.* **38**, 265–380 (1980).
23. VARANASI, U. and MALINS, D. C., 'Metabolism of petroleum hydrocarbons: accumulation and biotransformation in marine organisms', in *Effects of petroleum on Arctic and Subarctic marine environments and organisms*, Vol. II, (D. C. Malins, Ed.), Academic Press, New York: San Francisco: London, 1977, pp. 175–270.
24. MALINS, D. C., 'The chemical and biological degradation of petroleum: a foremost challenge for the analytical chemist', *Petroleum and the Marine Environment*, Graham and Trotman, London, 1981, pp. 319–344.
25. DICKS, B., 'The effects of refinery effluents: the case history of a saltmarsh', In *Marine ecology and oil pollution*, (J. M. Baker, Ed.) Applied Science Publishers, London, 1976, pp. 227–245.
26. LEMLIN, J. S., 'The value of ecological monitoring in the management of petroleum industry discharges: experiences in Esso Petroleum Company United Kingdom refineries', Proceedings of I.A.W.P.R. Conference on the Environmental Impact of Man's Use of Water, Brighton, 3–7 November 1980.

27. CONCAWE, *The Environmental Impact of Refinery Effluents*, Concawe Report No. 5/79, Concawe, Den Haag, 1979.
28. WILSON, K. W., 'Toxicity testing for ranking oils and oil dispersants', *Ecological aspects of toxicity testing of oils and dispersants*, Applied Science Publishers, London, 1974, pp. 11–22.
29. FROST, L. C., 'Torrey Canyon disaster: the persistent toxic effects of detergents on cliff-edge vegetation at the Lizard peninsula, Cornwall', *Cornish Studies* 2, 5–14 (1974).
30. GUZMAN, L. and CAMPODONICO, I. 'Studies after the Metula oil spill in the Straits of Magellan, Chile', *Petroleum and the Marine Environment*, Graham and Trotman, London, 1981, pp. 363–376.
31. DICKS, B., 'Some effects of Kuwait crude oil on the limpet, Patella vulgata', *Environmental Pollution* 5, 219–229 (1973).
32. BAKER, J. M., 'Ecological changes in Milford Haven, during its history as an oil port', in *Marine ecology and oil pollution* (J. M. Baker, Ed.). Applied Science Publishers, 1976, pp. 55–66.
33. STEPHENSON, T. A. and STEPHENSON, A., *Life Between Tidemarks on Rocky Shores*, W. H. Freeman, San Francisco, 1972, 425 pp.
34. MALLEY, D. F., HO SINN CHYE and CHARLES, J. K., 'Beaches, rocky shores and coastal swamps', in *Coastal resources of West Sabah: An Investigation into the Impact of Oil Spill*, (C. Thia-Eng and J. A. Mathias, Eds), Penerbit Universiti Sains Malaysia, Pulau, Pinang, 1978, pp. 152–170.
35. MACNAE, W., 'A general account of the fauna and flora of mangrove swamps and forests in the Indo-West Pacific region', *Adv.Mar. Biol.* 6, 73–270 (1968).
36. CHAPMAN, V. J. (Ed.), *Ecosystems of the World*, Vol. I, *Wet Coastal Ecosystems*, Elsevier, Amsterdam: London: New York, 1977, 428 pp.
37. BAKER, J. M., SURYOWINOTO, M., BROOKS, P., and ROWLANDS, S., 'Tropical marine ecosystems and the oil industry: with a description of a post-oil spill survey in Indonesian mangroves', *Petroleum and the Marine Environment*, Graham and Trotman, London, 1981, pp. 679–704.
38. PAGE, D. S., MAYO, D. W., COLLEY, J. F., SORENSON, E., GILLFILLAN, E. S. and HANSON, S. A., 'Hydrocarbon distribution and weathering characteristics at a tropical oil spill site', *Proceedings of the 1979 Oil Spill Conference*, API/EPA/USCG, 1979, pp. 700–712.
39. RAY, J. P., 'The effects of petroleum hydrocarbons on corals', *Petroleum and the Marine Environment* Graham and Trotman, London, 1981, pp. 705–726.
40. NICHOLS, D. (Ed.), *Monitoring the marine environment*, Institute of Biology, London, 1979, 205 pp.
41. BROWNELL, R. L., 'Whales, dolphins and oil pollution', in *Biological and oceanographical survey of the 'Santa Barbara' Channel oil spill 1969–1970*, (D. Straughan, Ed.), Allan Hancock Founda-

tion, University of Southern California, 1971, Chapter 12.
42. BROWNELL, R. T. and LE BOEUF, B. J., 'Californian sea lion mortality, natural or artefact?' in *Biological and oceanographic survey of the 'Santa Barbara' Channel Oil Spill 1969–1970*, (D. Straughan, Ed.) Allan Hancock Foundation, University of Southern California, 1971, Chapter 14.
43. GERACI, J. R. and SMITH, T. G., 'Consequences of oil fouling on marine mammals', in *Effects of petroleum on Arctic and Subarctic marine environments and organisms*, Vol. II, (D. C. Malins, Ed.), Academic Press, New York: San Francisco: London, 1977, pp. 399–410.
44. DAVIS, J. E., *Skomer Island National Nature Reserve Grey Seals (alichoerus grypus),* Warden's report on 1974 breeding season with special reference to the effects of oil pollution, 1974.
45. BOURNE, W. R. P., 'The Christos Bitas affair', *Marine Pollution Bulletin* 10 (5), (1979).
46. RICHARDSON, M. G., 'Esso Bernicia incident, Shetland', *Marine Pollution Bulletin* 10 (4), 97 (1979).
47. ELLIS, D. V., 'Sea otters at risk from oil spills', *Marine Pollution Bulletin* 10 (3), 68 (1979).
48. GUSEY, W. F. and MATURGO, Z. D., *Petroleum production and fish and wildlife resources in the Gulf of Mexico, Louisiana*, Environmental Conservation Dept., Shell Oil Co., New York, 1971.

4. Birds

Birds are particularly vulnerable to oil pollution because they encounter it floating on the water surface, and the more aquatic species in particular are accustomed to avoid hazards by diving, which causes them to become progressively embroiled with floating oil-slicks. It has been suggested that the more aerial species may also settle on slicks because they make the water calm, or resemble fish-shoals or other likely sources of food, though there seems to be little evidence to support this.[1] They certainly sometimes pick up floating tar-balls in mistake for food.[2]

If birds come into contact with liquid oil it soaks their plumage and destroys its waterproofing and insulating properties so that the birds are chilled and liable to die of exposure and exhaustion when their energy reserves are exhausted.[3] If they inhale or swallow toxic components of the oil while attempting to clean themselves this may lead to internal damage to the respiratory, alimentary or excretory systems.[4] When the oil has begun to lose its volatile and toxic components by evaporation and solution and begins to solidify it may no longer soak the plumage and cause internal damage if it is ingested, but the birds may still damage their plumage trying to clean themselves,[5] while if they smear oil on their eggs it blocks the pores in the shell through which respiration takes place so the embryo is poisoned and asphyxiated.[6,7]

Bird populations are particularly vulnerable to oil pollution because they are highly mobile, and large numbers of birds from a wide area are liable to gather together to feed, roost, breed or moult, or pass through vulnerable bottlenecks on migration. The damage may be very persistent because many pelagic seabirds in particular are long-lived with low reproductive rates, seldom breeding before they are several years old or rearing more than one chick per year, so that losses are not easily replaced.[8] Oil is therefore likely to cause more damage to bird populations than to any other form of wildlife if pollution is not controlled, and may already be limiting bird numbers in some highly-polluted areas.

Thus it has been suggested that the number of long-tailed ducks, *Clangula hyemalis* wintering in the Baltic, has been greatly reduced by oil pollution,[9] and that the vast populations of these and other wildfowl, moulting and wintering in its approaches, may soon be threatened.[10] The breeding populations of common eider, *Somateria mollissima*, in the Netherlands and the Gulf of St Lawrence area may also have been threatened by exceptional spills.[11,12] The alcids, including the large guillemots, razorbills, and smaller puffins, have decreased in numbers in the English Channel[13] and along the North shore of the Gulf of St Lawrence.[14] Pollution is probably one of the most important factors in the decline, though climatic changes may also be involved.

Mechanism of damage by oil to bird populations

Oil pollution is liable to trap birds in particular ways. If it is released into the upper reaches of a river, it is liable to spread down-stream and out through the estuary to the sea, catching in its progress the bird flocks feeding or roosting in the channels. If oil pollution is released or arrives at the mouth of an estuary or inlet with a rising tide, it is likely to be swept inland and again catch birds in the channels. If it is released after dark, it may catch birds asleep on the water, as occurred, for example in the autumn of 1966 when the *Seestern* accidentally pumped 1700 tons of oil into the River Medway and killed 2700 birds further up the estuary.[16] However, although the damage caused by the release of oil into an enclosed inlet may be limited, a particular risk is presented by a prolonged spill into an open strait, since oil is likely to be swept back and forth by the tide and cause pollution which spreads out to sea in both directions. The new oil terminals in Orkney and Shetland have had problems in this respect and otters have been killied as well as birds.

If a vessel starts to leak oil while under way at sea, it is likely to leave behind it a long trail which will drift with a beam wind to form a front which sweeps up all the birds ahead of it, trapping feeding flocks against the coast. This occurred in January 1967 when the *Saale* was damaged in a collision in the English Channel off Beachy Head and was towed, leaking, fifty miles down channel to Southampton Water, leaving a trail of oil which blew ashore, wiping out most of the seabirds along the coast of Sussex.[8] If a moored or wrecked vessel starts to leak oil as with the *Torrey Canyon* in the spring of 1967, it will first drift to lee to form another long trail, and then, if the direction of the wind changes, it will move before it as a front and again trap birds against the shore. During the passage of a series of weather systems the *Torrey Canyon* oil eventually brought over ten thousand birds ashore along both coasts of Cornwall and Brittany in this way (see Fig. 4.8) wiping out five-sixths of the French puffin, *Fratercula arctica*, population breeding on the Sept Iles in the process.[17] The damage was renewed with the loss of the *Amoco Cadiz* in 1978 and the *Tanio* in 1980.

Vulnerable bird concentrations are liable to be found in certain specific locations. Out at sea they occur in areas with a high marine productivity, around ice and the fronts between water-masses and along the boundaries of currents, in association with fish-shoals or fishing-boats which discard offal. Along the coast, they may gather to feed in, or around, the tide-races off headlands, in straits and the approaches to estuaries, over beds of shellfish or marine vegetation such as *Zostera* and *Enteromorpha*, and off sewage and effluent outfalls. The last may lead to the development of particular hazardous situations, as is the case where all Edinburgh's sewage was formerly discharged untreated into the southern Firth of Forth at Seafield, attracting up to 30,000 sea duck and other birds in the winter, immediately down-stream from the port of Leith and a new oil-terminal at Hound Point. Birds may be at risk from oil coming down the sewer as well as that discharged from docks and ships in such situations.

It should be noted that while floating oil drifts with the wind at about 3.3% of its speed, bird bodies only travel about two-thirds as fast, so that by the time they attract attention along the shore the oil may have been ashore for some time or drifted far away. On the other hand, surviving oiled birds may swim ahead of the oil to the nearest shore to get out of the water, or continue

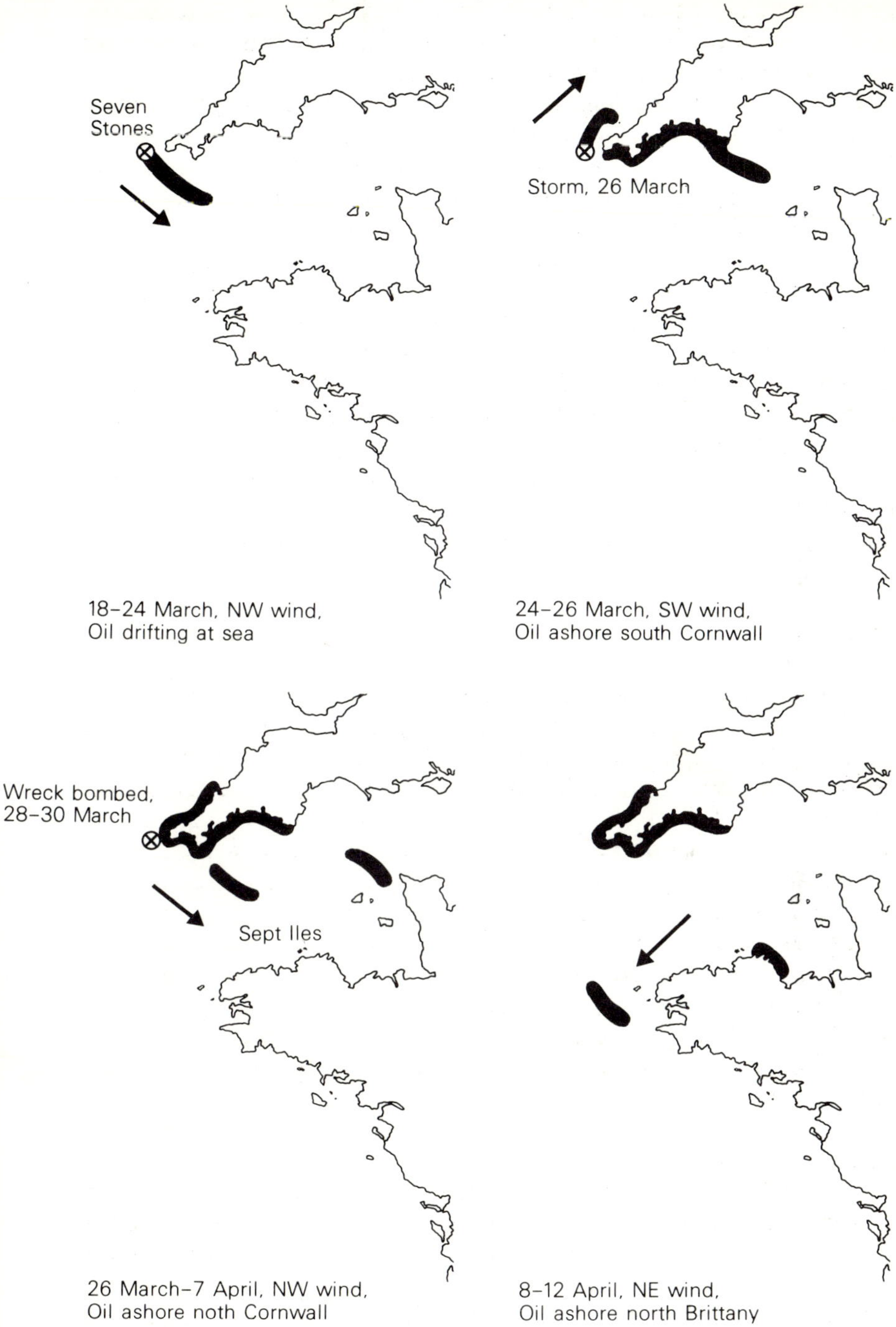

Fig. 4.8 Movements of the *Torrey Canyon* oil

established feeding-movements or migrations and appear far away. After about ten days, oiled bodies may sink and start to move with bottom currents, and yet subsequently be cast up again on the beach.[18] It may therefore be difficult to plot the distribution of pollution from that of bird bodies, and by the time they appear much damage may have been done.

The vulnerability of different types of birds

In general, most of the pelagic species, such as the albatrosses, fulmars, shearwaters and most other petrels, gannets, boobies, tropic-birds, frigate-birds, and some pelicans, skuas, gulls and terns, hunt on the wing, so although they may occasionally be caught by oil while resting on the water, or pick up tar-balls and oily objects in mistake for food or for nest-material, they do not suffer greatly from oil pollution out at sea. If they are affected, they may not be killed, as they are not so exposed to wetting and chilling as more aquatic species. They may be most vulnerable in breeding-places where up to half the birds present may be sitting on the water. They also tend to have low reproductive rates, so that damage may take a long time to be repaired.

The species which feed by swimming and diving are most vulnerable, including the penguins, diving-petrels, divers, grebes, cormorants, shags, swans, geese, duck, alcids and phalaropes. Some of them, such as the penguins and alcids (including the guillemots or murres, razorbill, puffins, little auk or dovekie, auklets and murrelets), occur far out to sea in high latitudes, but elsewhere they are more characteristic of coastal and enclosed waters, where they may gather to roost or feed in vast concentrations. The penguins cannot fly at all and come ashore to moult, but many others including the driving-petrels, divers, grebes, alcids and wildfowl (the swans, geese and ducks) lose all their flight-feathers simultaneously as they moult and take refuge on the water. Birds from vast areas may then gather together at a good feeding-ground offshore, as for example where most of the north-west European shelduck, *Tadorna tadorna*, congregate in the Heligoland Bight, so that the whole population becomes very vulnerable. However, some of these birds have comparatively high reproductive rates which may account for their survival.

Birds which feed along the shore or inland, either from the air or on the ground, such as the gulls, waders or shorebirds and some wildfowl, are usually able to pass over oil pollution without

Fig. 4.9 Guillemots on the sea

getting too dirty, and survive well even if they become conspicuously polluted. They are most vulnerable when they roost on the water, whether by night, like the gulls, or by day, like some night-feeding wildfowl. They are liable to be affected by disturbance of their breeding places during incidents and may desert them or get their eggs oiled and they tend to have rather high reproductive rates.

In addition to variations in their vulnerability and recuperative rate, birds may also respond differently to attempts to rehabilitate them when oiled. In general, individuals of most species can be maintained for a time in captivity if they are cared for with sufficient skill, absolute cleanliness being an important factor[19] (see Appendix 1), but the results when inexperienced people with inadequate facilities try to deal with large numbers of birds in an emergency tend to be less good. The useful contribution to the survival of a species or population made by rehabilitating and releasing birds also varies. Thus if millions of birds such as alcids or penguins are becoming oiled, it may be less effective to rehabilitate a few of them than to attempt to preserve a few individuals of scarce and attractive species, such as the grebes and divers, also exposed to other pressures.

Areas at risk and precautions against damage

The damage caused by oil pollution to bird populations appears to be very much worse in cool climates during the winter, where the oil tends to persist in a toxic liquid state, than in warm climates and during the summer, where it is soon converted into inert solid residues, comparatively innocuous to birds. In consequence, the effects of oil pollution tend to be worse in hard winters, as in 1969–70 on the east coast of Britain, and also all the down the west coast of Europe following many storms in 1980–81.

If we consider the world as a whole, the largest and most vulnerable bird populations, including many swimming species, are found at sea in high latitudes, especially around the ice in both hemispheres, and also where biologically rich cold currents set into low latitudes along the continental coasts, off northeast Asia, northeast North America, southeast and western South America, California, west and southwest Africa, southeast Australia and New Zealand; and also along the southern coasts of Arabia which are influenced by the monsoons. Few or no swimming species, vulnerable to oil pollution, occur far out to sea in middle latitudes, and especially in the tropics, though large numbers may occur in enclosed waters such as the Gulf of California and the Persian/Arabian Gulf. Such places may also provide staging-posts for vast populations of migrants.

By far the best and most economical defence for bird populations is clearly to avoid pollution. Once oil is widespread in areas with large bird populations the chance of remedial action is normally low, and it may be difficult and expensive, especially in darkness and bad weather. The best hope from the ornithological point of view is usually that the oil may be immobilised as soon as possible by coming ashore on some biologically uninteresting stretch of coast.

It is possible to postulate some potentially useful actions, however. There may be little ornithological justification for attempts to control pollution far out to sea before its destination becomes clear, but if it is found to be heading for bird concentrations the sooner attempts are made to isolate or disperse the relevant slicks or prepare defences against their arrival, the better. If this cannot be done, it may sometimes be possible to scare the birds away before the slick arrives. Such an operation would have to be planned and timed very carefully to remove the birds soon enough, frighten them far enough away, and guard against their premature return before the oil has passed on. It might be achieved with boats, hovercraft, helicopters, or low-flying aircraft. It might prove difficult to ensure that floating bird-scarers moved with the oil, though they might work for a time if they were moored in vulnerable areas; birds soon become habituated to alarming stimuli, however, and the effect of the bird-scarers might not last long. With persistent action there is probably a good chance that winter flocks of wildfowl might be broken up and driven away from threatened areas (wildfowlers do it regularly), and a poorer chance that birds might be kept out of the path of a small oil-slick passing a large breeding colony. It might also be possible to catch breeding or moulting birds and keep them in captivity until the threat has passed by.[22]

While attempts at mass rehabilitation have rarely been very successful, and are best with penguins,[15] if even a few birds from small threatened populations can be saved it may make an important difference to their chance of survival. The view has been expressed[20] that it might be better to try to start new breeding colonies in safe places with such birds than to return them to their place of origin to get into trouble again. A growing number of ringing or banding recoveries now indicate that in fact such birds are usually very faithful to their natal colony and are likely to try to return there, so this may not be very effective unless a second generation of birds can be bred in captivity in the way pioneered by Marsault[21] and released at the site where it is hoped to start a new colony. It might be possible to re-establish some lost colonies in this way as well, because once the native stock has been exterminated it may take the birds a very long time to re-establish themselves. They are much more likely to do so if there are already birds in the vicinity, even if they are kept in captivity. There is also a need for more protection for breeding colonies to enable populations reduced by pollution to replace their losses.

References

1. BOURNE, W. R. P., in *Sea Birds and pollution*, (R. Johnston, Ed.) Academic Press, London, 1976, pp. 403–502.
2. DIAMOND, A. W., *Seychelles. Ann.Rep.*, International Council for Bird Preservation (British Section) for 1973, 1974, 23pp.
3. HARTUNG, R., 'Energy metabolism in oil-covered ducks', *J. Wildlife Manager* **31**, 798–804 (1967).
4. CLARK, R. B. and KENNEDY, R. J., *Rehabilitation of oiled seabirds*. Department of Zoology, University of Newcastle-upon-Tyne, 1968, 57 pp.
5. BOURNE, W. R. P. and BIBBY, C. J., 'Temperature and the seasonal and geographical occurrence of oiled birds on west European beaches', *Mar. Poll.Bull.* **6**, 77–80 (1975).

6. RITTINGHAUS, H., 'On the 'indirect' spread of oil pollution in a seabird sanctuary', *Orn.Mitt.* 8, 43–46 (1956) (in German).
7. HOLMES, W. N. and CRONSHAW, J., 'Biological effects of petroleum on marine birds', in *Effects of Petroleum on Arctic and Sub-Arctic Marine Environments and Organisms*, Vol. 2, *Biological Effects*, (D. C. Malins, Ed.), Academic Press, New York, 1977, pp. 359–398.
8. BOURNE, W. R. P., 'Oil pollution and bird populations', *Fld.Std.* 2 supp., 99–121 (1968).
9. LEMMETYINEN, R., 'Damage to waterfowl caused by waste oil in the Baltic area', *Suomen Riista* 19, 63–71 (1966) (Finnish with English summary).
10. JOENSEN, A. H., 'Danish seabird disasters in 1972', *Mar.Poll.Bull.* 4, 117–118 (1973).
11. SWENNEN, D. C. and SPAANS, A. L., 'Seabird mortality by oil in the Wadden area in February 1969', *Het Vogeljaar* **18**, 233–245 (1970) (Dutch with English summary).
12. BROWN, R. G. B., GILLESPIE, D. I., LOCK, A. R., PEARCE, P. A. and WATSON, G. A., 'Bird mortality from oil slicks off eastern Canada, February–April 1970', *Canad. Field Nat.* **87**, 225–234 (1973).
13. CRAMP, S., BOURNE, W. R. P. and SAUNDERS, D., *The Seabirds of Britain and Ireland*, Collins, London, 1974.
14. NETTLESHIP, D. N. and LOCK, A. R., 'Tenth census of seabirds in the sanctuaries of the north shore of the Gulf of St Lawrence', *Canad. Field Naturalist* **87**, 395–402 (1973).
15. WESTPHAL, A. and ROWAN, M. K., 'Some observations of the effects of oil on the Jackass penguin', *Ostrich Supp.* **8**, 521–526 (1970).
16. HARRISON, J. and BUCK, W. F. A., *Peril in perspective: An account of oil pollution on the Medway estuary*, Special supplement to the *Kent Bird Report* Vol. 16. Kent Ornithological Society, 1967 17 pp.
17. BOURNE, W. R. P., 'Special review: after the Torrey Canyon disaster', *Ibis* **112**, 120–125 (1970).
18. HOPE JONES, P., HOWELLS, G., REES, E. I. S. and WILSON, J., 'Effect of Hamilton Trader oil on birds in the Irish Sea in May 1969', *Brit. Birds* **63**, 97–110 (1970).
19. Advisory Committee on Oil Pollution of the Sea, Research Unit on the Rehabilitation of Oiled Seabirds, *Recommended Treatment of Oiled Seabirds*. Department of Zoology, University of Newcastle-upon-Tyne, 1972.
20. BOURNE, W. R. P., 'Oil pollution and bird conservation', *Biological Conservation* **2**, 300–302 (1970).
21. MARSAULT, B. M., 'Auks breeding in captivity', *Bird Study* **22**, 44–46 (1975).
22. KOSKI, W. R. and RICHARDSON, W. J., *Review of waterbird deterrent and dispersal systems for oil spills*, PACE Report No. 76–

6, Petroleum Association for Conservation of the Canadian Environment, 130 Albert St, Ottawa, Ontario, 1976, xviii + 122 pp.

Chronological abstracts and an alphabetical bibliography of references to oil pollution affecting birds will be found in Vermeer, R. and Vermeer, K., *Oil pollution of birds: an abstracted bibliography*, Pesticide Section, Canadian Wildlife Service, Manuscript Report No. 29, 68 pp., 1974, obtainable from Wildlife Toxicology Division, Canadian Wildlife Service, Ottawa, Ontario, Canada K1A 0E7.

Part Two

Treatment of Oil on the Water

5

The use of booms

Because oil spilt on water spreads naturally it is desirable to try to confine a spill to a point near its source. Clearly, oil spilled in a dock or harbour basin may be confined by the walls of the basin, but it will also be concentrated, to some extent, by wind blowing over the area, and individual pockets will sometimes be formed. If the total area over which the oil has spread can somehow be reduced, then its thickness will be increased, and there will be better chances of physical recovery. Everyone now agrees that the physical removal of the oil is to be preferred whenever possible.

It is wishful thinking to expect that oil may be recovered from the water surface under any conditions. If this were so, it would be possible for considerable amounts of oil to be salvaged. Unfortunately, the areas where successful containment of oil on the water surface can be achieved are limited by wind, wave and current conditions to the relatively calm waters, usually only found in some rivers and estuaries. Recovery in the open sea, is possible only in very good conditions. There are two main equipment requirements—a boom to retain the oil, and some method of picking up the oil.

Oil booms are floating barriers, used to prevent the spreading of oil on water, and sometimes to thicken the layer of oil by reducing the area into which it has spread. In its simplest form, an oil boom may be described as a floating fence supported by one or more buoyancy units. There are many variations in design and attempts have been made to elaborate the system by incorporating weirs to skim off, and so remove, the entrapped oil.

Types of booms

Commercially-made booms are usually one of two types called 'fence' or 'curtain' booms (see Figs 5.1 and 5.2).

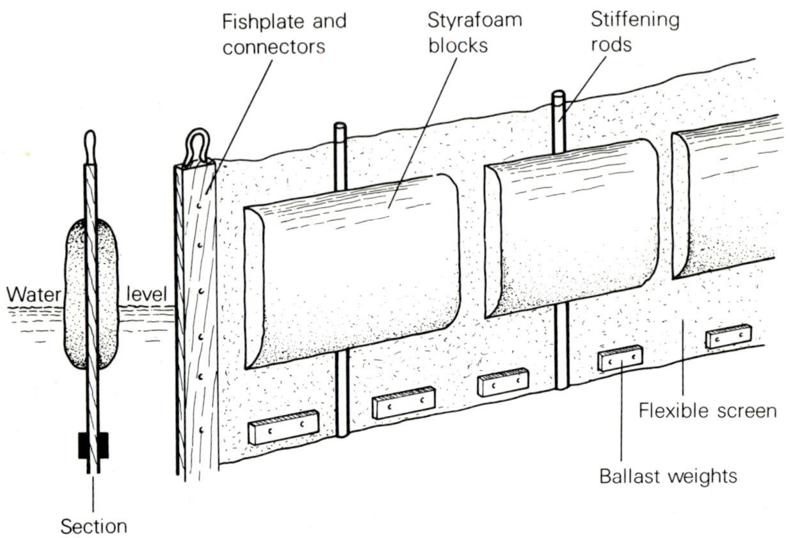

Fig. 5.1 Typical 'fence' type boom

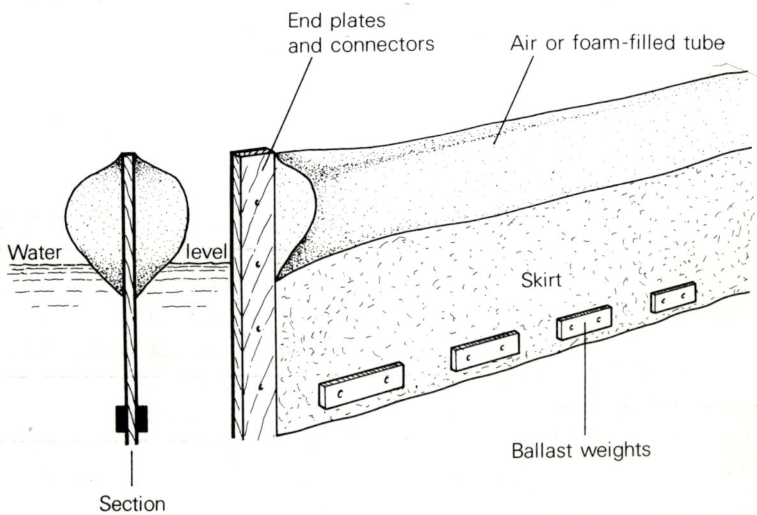

Fig. 5.2 Typical 'curtain' type boom

The fence boom is one made of rigid or semi-rigid material, presenting a vertical screen against the oil floating on the water. The screen is kept afloat by plastic-filled buoyant compartments or blocks of plastic foam and is usually ballasted by weights attached to the base of the screen. (Examples of this type of boom are made by Petro Barrier and Simplex*.) Some manufacturers offer booms of this type in 'kit' form in which the main component, or screen, may be rolled-up and compactly stored. The other parts are contained in sacks, which means that the floats and ballast weights may have to be assembled on the site. It can be seen therefore, that in order to obtain compact storage and ease of transport, this type of boom's readiness for immediate use has to be sacrificed. There are many types of boom in this category.

The other common type of boom, the curtain design, is made up of long and usually continuous buoyancy tubes carrying a pendant skirt with chains or metal weights attached to the base. (Examples of this type are the Warne and Kepner booms.†) Often such booms are air-inflated before deployment and deflated after use. Thus bulk in storage is saved, but some sort of inflation device has to be carried as additional equipment. However it is probably easier to inflate this type of boom than to assemble the many parts of which some 'fence' type booms are composed. There are also other reasons why the curtain type boom is to be preferred.

In order to keep the skirt down in high current conditions, the ballast weight must be increased and this must be matched by an increase in buoyancy. If the buoyancy tube is of the 'layflat' variety more air can be pumped in to restore the initial level of flotation. The submerged portion of a boom may be subjected to high drag forces due to currents; these forces are proportional to the square of the velocity (v^2).

The Hydraulics Research Station at Wallingford, England, has carried out field tests on a number of booms of different types and it is clear from the results that the performance of vertical screens is far below that of some of the heavier curtain type booms.[1] Apart from the increased windage of most vertical fence booms, they rarely have sufficient buoyancy to overcome their tendency to dive below the water surface in currents (see Fig. 5.3a) or, if moored from a point near the waterline, have sufficient stability to remain vertical (see Fig. 5.3b). The performance of a curtain boom under similar conditions is much more satisfactory (see Fig. 5.4).

*Crowley Environmental Services Corporation, San Francisco CA. American Marine Inc. Florida.

†William Warne & Co., Barking, London. Kepner Plastics, Torrance, California.

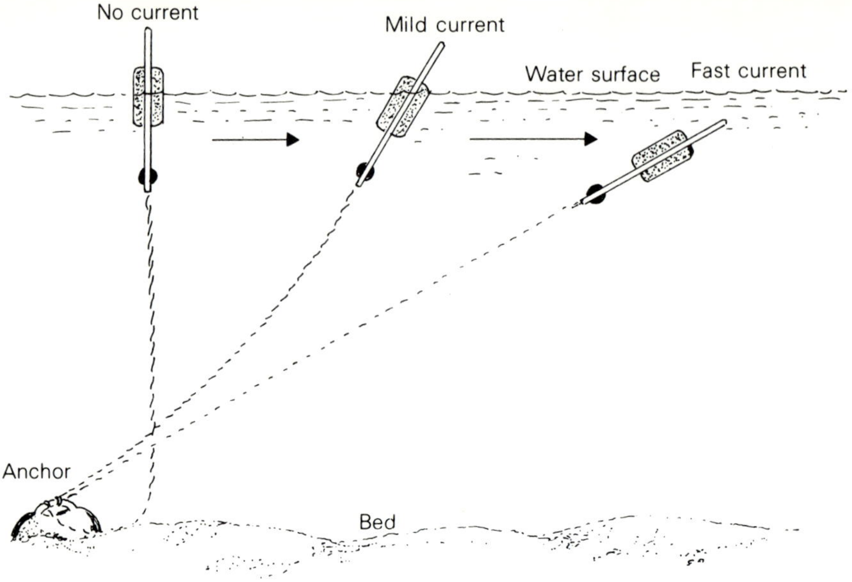

(a) Mooring line attached at base

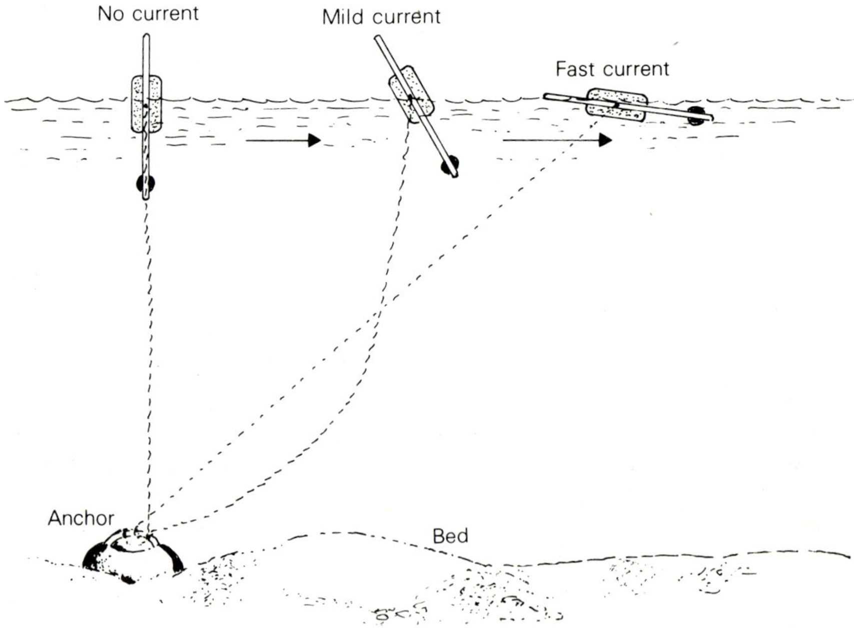

(b) Mooring line attached at waterline

Fig. 5.3 'Fence' type boom (a) mooring line attached at base; (b) mooring line attached at water line

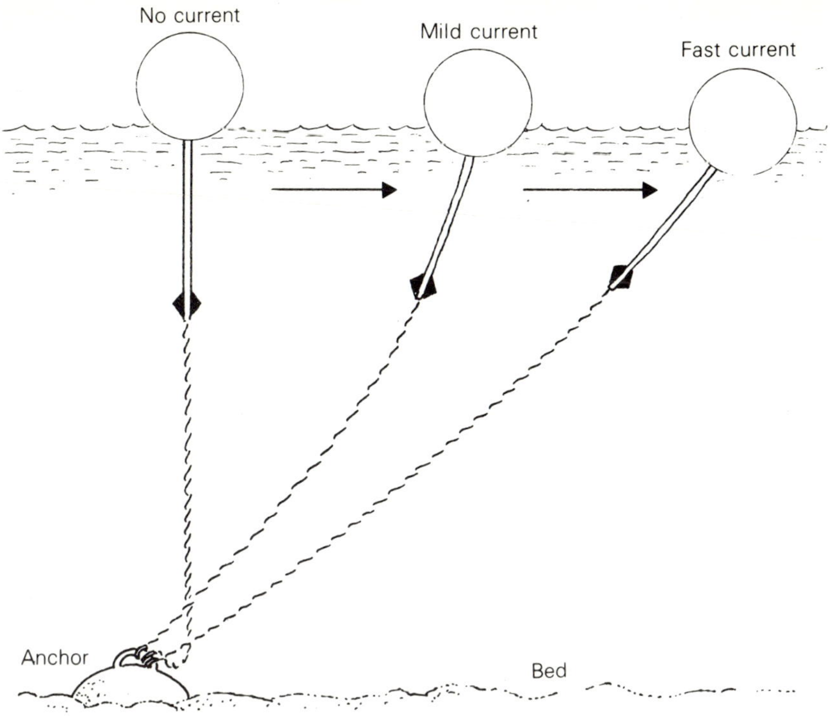

Fig. 5.4 'Curtain' boom moored at base

The two types of boom outlined above constitute at least 80% of all known commercial types. The remainder consist of a few composite designs, some multitubular in section, and others that have the profile of a 'curtain' boom, but with the buoyancy portion made up of hemispheres of foamed or other types of plastic. The skirt is usually rigid or of flexible material held in place with spines or stiffeners.

At least one make of boom is an all-metal product. For example the type made by Gamlen-Naintre & Cie* has rectangular aluminium floats on both sides of a vertical metal screen and is produced in short sections joined by chains and reinforced flexible connectors. This is a fire resistant boom which has been found quite effective for this purpose though not ideal in currents much in excess of a knot (0.51 m/s). The reason for this is that the individual floats give rise to turbulent flow and the eddies formed at the ends of the floats cause the surface flow to dive beneath the boom, taking the oil with it.

A few types of boom work on the principle of absorbing the oil into packages of oleophilic crumb. The crumbs of plastic may

*Gamlen Europa SA Marly-le-Roi, France

absorb up to twenty times their own weight of oil, and are contained in net sleeves or other permeable containers. Clearly their oil absorbing properties are finite and as each boom becomes saturated with oil it is replaced by a new length. The booms are therefore disposable and although the cleaning of soiled equipment is avoided, they are not easy to handle, as overcompression of the plastic causes some loss of oil. Consequently these absorbing booms are of little value except for limited spills on enclosed waters where complete removal of the floating oil is required, e.g. on open reservoirs used for domestic water supplies. One example of this type of boom is made by SOSOL.* A simplified form of a boom of this type may be made up in an emergency by wrapping quantities of loose straw in a fishing net, but this will have a reduced capacity for absorption, and sinks after prolonged use, although silicone treated straw gives better results.

Another device which has recently appeared on the market is the Oil Mop. It consists of a rope made of oleophilic fibres carefully woven into a polypropolene core thus forming an efficient oil absorbing device with a density of 0.9. Although primarily designed as a pick-up device, intended to be passed between rollers to expel the oil, it may also be described as a floating absorbent boom, and the principle could be applied to the removal of oil from small streams, accidentally polluted, possibly from a factory discharge. Fig. 5.5 shows a simple layout that could be applied to such a case.

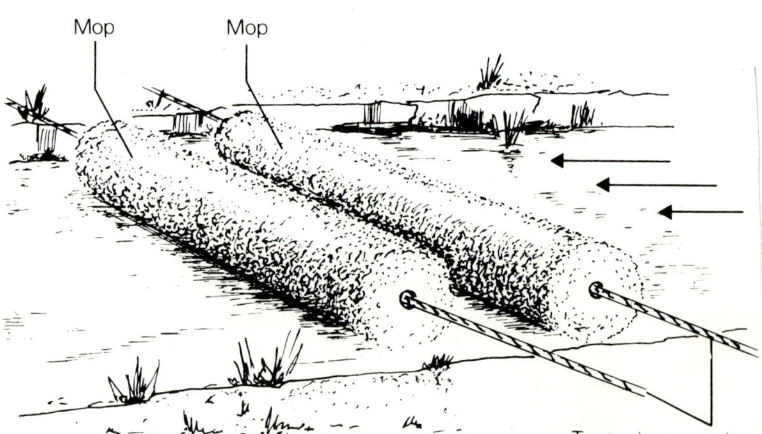

Fig. 5.5 Cylindrical mops of oleophilic fibres

*SOSOL. Axel Christiernsson, Gothenberg, Sweden.

One mop is placed across the stream and rotates slowly in the flow. Before the dirty mop is removed, a second one is placed in position further upstream. If the pollution was recurrent then the complete Oil Mop Machine (see p. 137) might be installed.

An important consideration in boom design is to produce a long smooth barrier without too many protruding appendages. Once again a compromise has to be reached between packing and stowing—and hydraulic design. For example, long unbroken lengths of tubular boom (see Fig. 5.6a) present less resistance to flow than shorter lengths joined by some means that is convenient to operate but usually forms an expansion or contraction, which in itself hinders the smooth flow along it (see Fig. 5.6b). If the boom, however, is one long continuous length it is unwieldly to transport, handle and repack and it must be broken into sections having a smooth connecting piece at the joints between sections (see Fig. 5.6c).

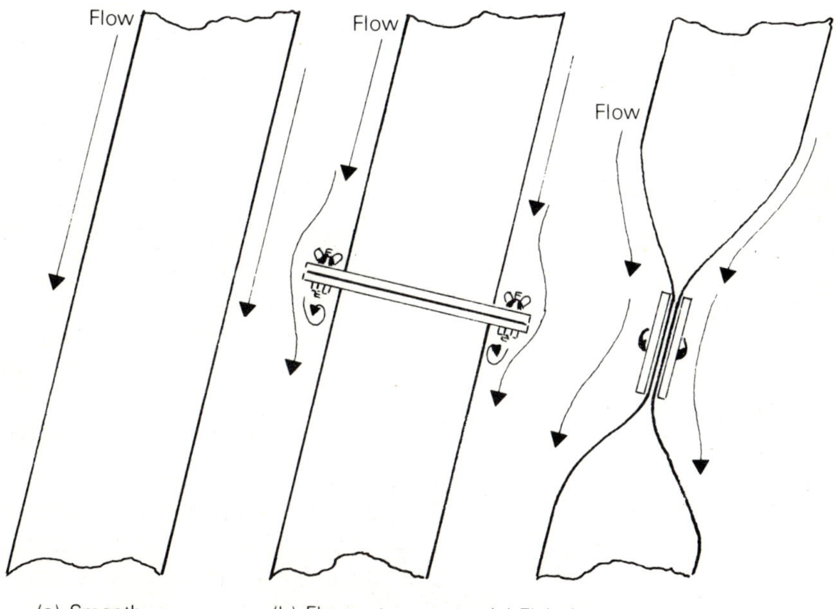

Fig. 5.6 Plan form of booms – connectors

Most booms are made of reinforced plastic and some are proofed against the effects of sunlight (hyperlon coating) and oil. Those made from nylon-reinforced neoprene seem to have the most durable qualities. Long storage life, even when oily, is an essential requirement.

The strength of the material is all-important in boom design, but very often one finds that material of adequate strength has been weakened by the addition of inadequate fittings. A boom 300m long, moored in tidal flow, might have to stand several tonnes load on its end moorings. If this load is transmitted to the mooring throughout the full width of the boom material there is usually ample strength, but if the points of attachment are reduced to a few separate eye fastenings, then the load is concentrated at the weakest points and failure often occurs.

Other booms have built-in load-carrying members, such as parafil or wire ropes to which the end-plates of the boom are attached. This is often a satisfactory method, but requires additional material resulting in added expense.

At a few harbours and oil terminals, rising and sinking booms have been installed.[2] These types of boom are usually of the 'fence' type and may be inflated and deflated as required. The obvious advantage of this system is the ease with which they can be removed from entrances and navigational channels. A main disadvantage is their liability to damage through careless use of anchors and mooring lines from ships using the port.

A boom of a rather different type has been developed by BP Ltd. The flotation element is an air filled tube but the skirt is provided by a second tube filled with water. There is a third smaller tube on the top of the other two. The structure and dimensions are shown in Fig. 5.7. The three tube configuration gives the boom effective draught and freeboard even under severe sea conditions, making it extremely flexible and manoeverable and gives it good wave following characteristics. The small upper tube, or inflation cuff, prevents the boom from twisting during the laying operation and aids recovery after use.

The particular advantage is that, uninflated, the boom lies flat, and so can be stored in a small space. One arrangement, called

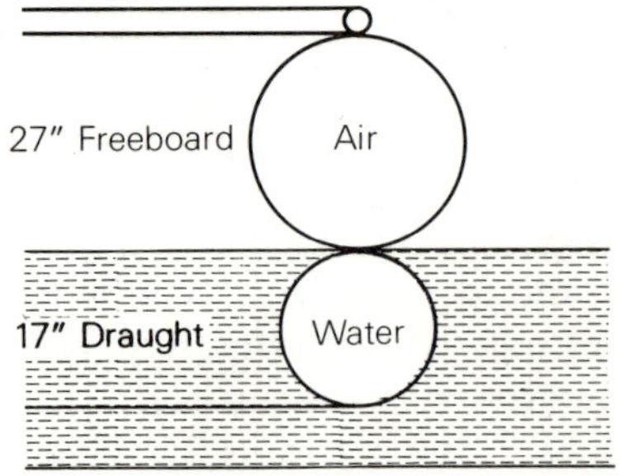

Fig. 5.7 The Vikoma boom

the Vikoma Seapack, consists of a small glass reinforced plastic boat, 23 ft long which contains 1475 ft of boom. Within the Seapack hull is also a diesel driven fan and a ducted propellor water pump which simultaneously inflate both the air and water sections of the boom to which they are permanently connected. The cuff is inflated from compressed air cylinders which are also carried in the Seapack.

Seapack may be launched from a slipway, or lowered into the water by a crane. Thereafter the diesel engine is started and it is towed to the operation site. No crew is required nor manual attention after launching until it is recovered after use. Boom deployment is initiated either by launching a small deployment drogue, already attached to the free and of the boom, or by attaching the end to a suitable boat or to an anchor or pier if in confined waters. The Seapack is then towed away at 3–4 knots and its 1475 ft of boom lays and inflates automatically in some 15 minutes.

Deployed down wind of a slick, holding one end with the towing boat and the other with the drogue, or preferably with another boat, the boom will collect and concentrate and contain the oil (Fig. 5.8). Its particular advantages are the ease of deployment with a very limited crew.

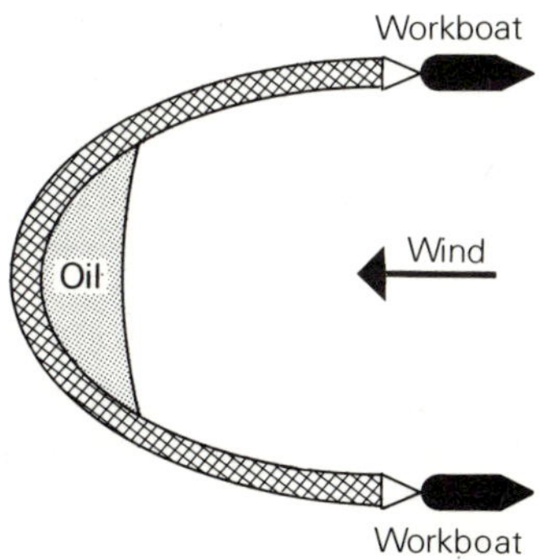

Fig. 5.8 Vikoma in operation

A larger alternative intended for operation in rough seas and high winds is the Vikoma Ocean Pack. This has a similar boom made of a heavier fabric and without the inflation cuff. The 500m of boom complete with power unit, a gantry, power-operated loading capstan, and operator cab comprises a complete unit 5.18 m long 2.41 m wide and 2.5 m high (with gantry stowed) weighing 5.8 tonnes. This unit can be quickly fitted to a variety of vessels.

Fig. 5.8a Deployment of Vikoma boom from Seapack

Air barriers

Air barriers also known as air curtains, bubble barriers etc. are used in some circumstances to prevent oil from reaching certain stretches of water. Initially devised for use as a wave breaker, they were tried as an oil barrier in 1937. Subsequently a good deal of work has been carried out.

An air barrier, as its name implies, is a curtain of air bubbles generated on the bed of the sea or channel and stretching upwards to the surface. To form the bubbles, a pipe or manifold suitably perforated is laid either on the bed, or in a shallow trench, and air is supplied by a suitable compressor.[3]

The upward flow of air bubbles generates an upward flow of water which, on reaching the surface, spreads out as two horizontal currents at right angles to the line of bubbles. In still, or nearly still, water the air bubbles form a plume as an inverted cone structure with a hump or ridge on the surface from which the water flows away on both sides (Fig. 5.9). This local current prevents the oil from crossing the barrier. This depends on the amount of air in the air/water cone below the surface. This is obtained by placing the air pipe as low as possible and using small bubbles to keep the rising bubble velocity as small as possible.

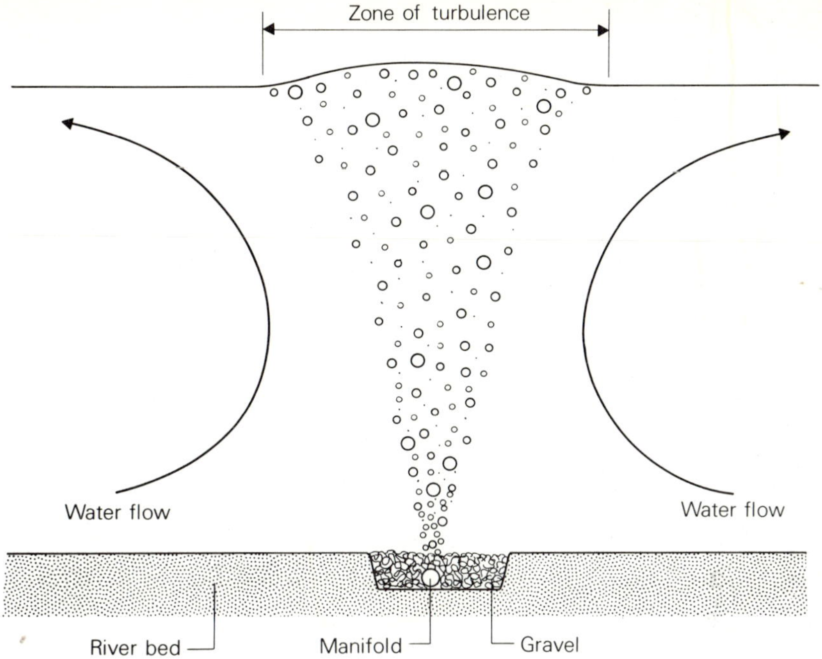

Fig. 5.9 Air barrier under ideal conditions

The bottom pipe is subject to damage from ship's anchors etc. but this can be avoided by laying the air pipe in a shallow trench across the water way and covering it with coarse gravel.

Installations of this type of between 300 m and 1000 m in length have been constructed. The amount of air varies between 0.2 and 1.3 m^3/m run of pipe. The pressure will depend on the depth of the manifold. Holes are spaced some 300 mm apart and vary in diameter from 0.5 mm to 0.7 mm according to the depth, the smaller holes being used at the greater depth.

Under the ideal conditions described above the air barrier will retain most of the spilt oil, when used to protect still water. Unfortunately this does not hold where there is an appreciable current, this upsets the air bubble pattern. Water currents deflect the rising plume, the height of the ridge is smaller and as a consequence the surface water flow is much reduced. This is shown in Fig. 5.10. Oil that approaches the barriers is swept away from it by the current giving a clear or 'stagnation' zone. The current can tear droplets from the accumulated oil, as in the case of a normal boom. Some of these are swept into the air plume and then move with the surface current on to the 'wrong side' of the barrier thus transferring the oil. Currents as low as 0.75 ft/s are sufficient to carry oil across the barrier. If air barriers are used in conjunction with efficient skimming devices, and if the barrier can be placed at a suitable angle to the current they can be made to operate reasonably well.

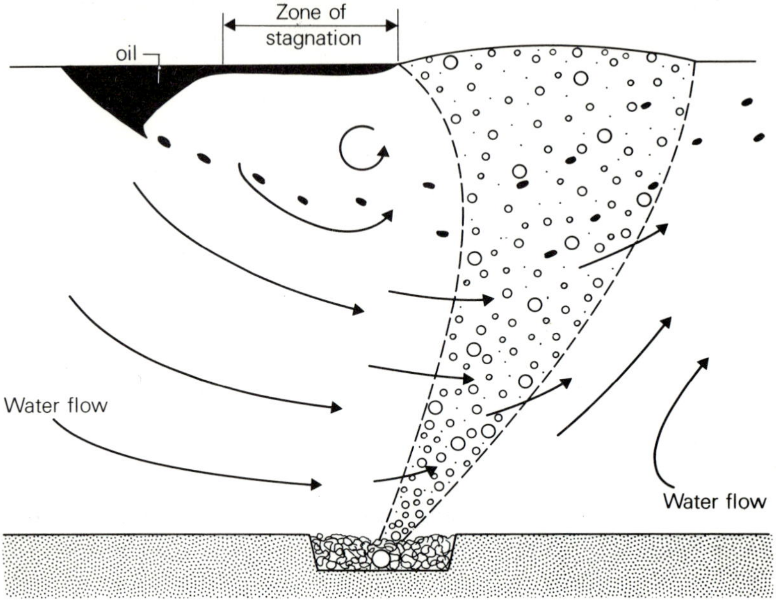

Fig. 5.10 Air barrier operating in current

They have been developed to form a permanent installation. Once installed they can be left inoperative until the occasion arises. Their big advantage is that they do not obstruct the water way to the passage of shipping. However, the provision of a special compressor, and its routine maintenance, can make them expensive to install and operate.

Surfactant chemicals

One further method of concentrating oil spilled on water is the use of surface-active chemicals. If one of these is discharged on to the water surface it rapidly spreads and the pressure exerted by the film can exceed the spreading pressure of an oil. The oil layer is then forced back by the chemical. Clearly such chemicals need to be insoluble in water and should not emulsify the oil, which would then becomes more difficult to remove from the water surface. Their toxicity to living creatures should also be minimal.

There are now several of these chemicals on the market among them are 'Oil Herder' the first one developed by Shell and B.P.'s 'Oil Marshal'. They have been used successfully, but as their maximum life in calm waters is about 24 hours, they can only be considered as a temporary method. In calm waters it would appear that the reduction in size of large slicks, which have been surrounded by one of these chemicals, could be followed by a more permanent method in which the concentrated oil is corralled physically by a suitable boom. By this means, only a limited amount of boom would be required. It is understood that approximately 12 gallons of 'Oil Herder' would be necessary to treat a slick of 1 mile in perimeter.

For a very large slick, aerial spraying from a helicopter or slow flying aircraft may be considered. In this case the rate of application should be doubled, to allow for losses by wind drift. In both cases spray nozzles should be chosen such that a uniform spray pattern, similar to normal rainfall, is produced.

Field tests on these surface-film forming chemicals have been carried out in the USA[5,6] and in the Netherlands.[7] In one test 25 gallons of oil were dumped from a boat some seven miles offshore in the Pacific Ocean. The spill was encircled by a fast boat injecting the chemical from a hand-held pump oilcan. It is stated that only 176 ml of chemical was used to obtain maximum concentration of the oil. Although a three-to-four foot swell was present at the time, 45 minutes had elapsed before the oil began to respread.

The Dutch experiments covered several possible alternative uses of 'Herder'. At sea, a number of artificial oil slicks were created and later 'contained' by the application of 'Herder' around them.

In a dock, tests were carried out to ascertain the effectiveness of 'Herder' in water already polluted by dissolved chemicals. Again, the results showed that the surfactant layer confined the oil, which was no longer able to spread.

It is generally conceded that chemical surface-forming films may be used to condense an oil spread provided that its final thickness does not exceed 0.5–1.0 cm. It is thus a good method for reducing large area spreads of small thickness, particularly in harbours, where the oil may be later collected. It appears that these monolayers will not operate if opposed by a wind strength of greater than 4 knots. The chemical can be used to prevent the pollution of beaches if sprayed on the water surface between the oil slick and the shore provided that the wind is not onshore. The use of this chemical on the beach is discussed on p. 212.

The Weir Boom

It can be seen that a well designed and deployed boom is highly effective in collecting oil from a thin layer into a smaller, but thicker, area. The problem is how to pick up and remove the thicker layer. This is discussed in detail in the next chapter. One method, depending on a boom, can be discussed here. It had long been believed that, if some form of aperture, at the water level, could be made in the boom itself, then the oil would enter it, and so could be removed.

A few years ago in the USA, Ocean Systems Limited developed an ingenious system, incorporating both primary and secondary weirs, to separate the oil from the water (See Fig. 5.11). Individual units, each about 1.25 m long are joined together to form booms of up to 30 ft in length, that are towed slowly towards the oil slick. The oil, together with some water, passes over a primary weir into a compartment made of an oil-resistant fabric which provides a quiescent area in which waves and water currents are much reduced. A secondary weir then allows the oil to accumulate in a second compartment, from which it may be pumped away.

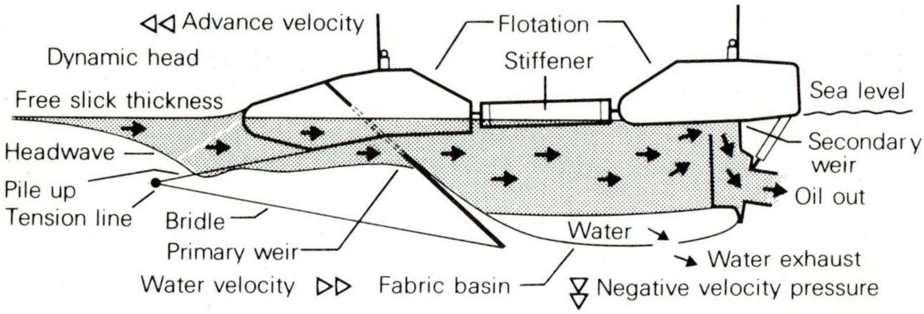

A cross-sectional diagram of the weir section modules

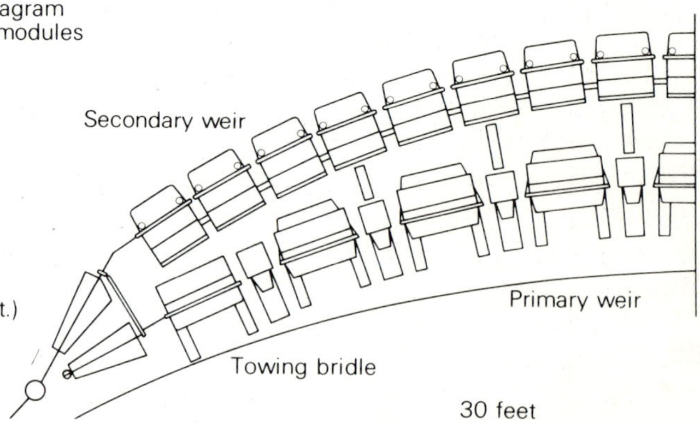

Plan view showing half of the modules of ORS-2000 deployed 0.61 m (2ft.) above water line; 0.61 m (2ft.) below water line

Fig. 5.11 Ocean Systems Barrier

Considerable attention has been paid to the design and shape of the various sections and of the flexible bottom of the first compartment area to ensure that, as the whole system is gently towed through the water and the oil slick, an area of negative pressure is produced beneath the bottom of the first compartment, so that the water below the oil in this compartment flows gently through holes in the flexible membrane.

In full operation, it is intended that each unit shall be fitted with its own pump and also with oil level detector switches, so that the pump operates only between certain limits of oil depth in the second compartment. In practice, the pumps would feed a common manifold, transferring the oil to a single oil storage vessel.

Field trials showed certain limitations and it is believed that the project has been dropped.

In the United Kingdom, The British Petroleum Company Limited, with financial support from the Norwegian and British Governments have recently completed their development of a weir boom system to collect and recover oil in a single operation.[8] The boom has a 4-tube configuration (Figs 5.12 and 5.13). A series of weirs is built in along the boom between the main air flotation tube and the water-filled skirt tube. Oil and water, or mousse and water, flowing across the weirs into the channel formed by the 4 tubes, are pumped away along the discharge tube by pumps situated one behind each weir, to the deploying vessel, where it is raised to deck level by a separate suction pump. The second oil/water mixture can then be discharged to intermediate storage.

The specially designed pumps in the weir boom discharge tube are hydraulically operated, the hydraulic lines being located in the smaller air buoyancy tube above the discharge tube.

A full scale research prototype of this system incorporating 10 weirs, each about 4 ft in length, was tested in late 1979 at the IXTOC 1 blow out site in the Mexican Gulf. A Vikoma Oceanpack was used to deflect oil into the operating area of weir boom array.

The most severe conditions encountered during this sea trial were wind speed 21–26 knots, with a 2 m swell topped by 0.7 m waves. Operating at $^2/_3$ capacity the system collected 500 t/h of a 40/60 mixture of mousse/seawater.

It also was shown to be capable of collecting mousse with virtually no free water under certain operating conditions, though at a lower overall liquid recovery rate of about 300 t/h.

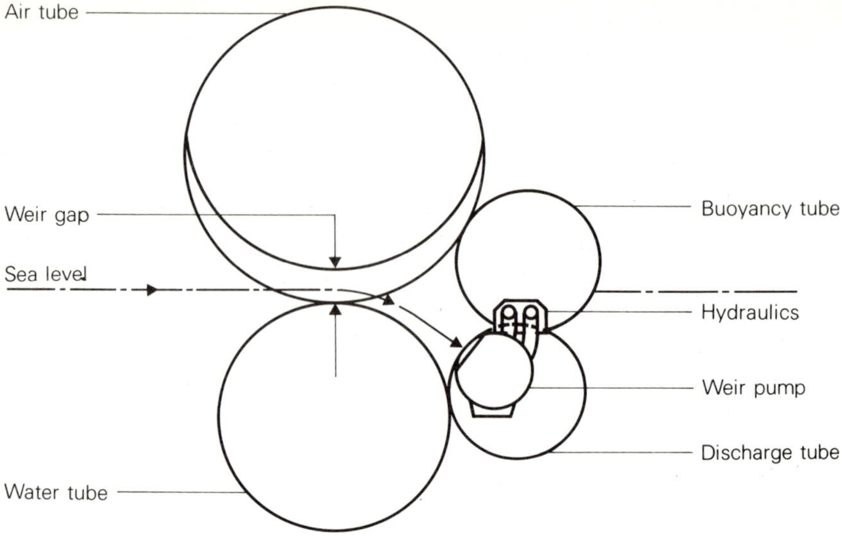

Fig. 5.12 Weir boom, cross-section

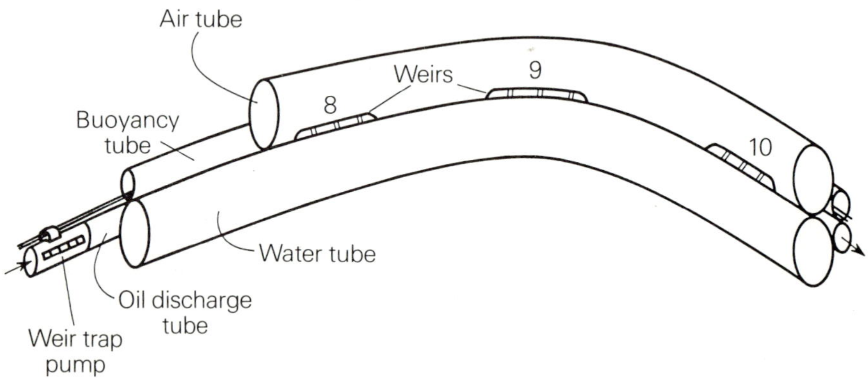

Fig. 5.13 Weir boom, construction

Availability of oil booms

There are very many firms specialising in the manufacture of oil booms, in many American and European countries as well as in

Japan. As booms are devices which can be made by manufacturers of many rubber and plastic articles e.g. hose and inflatables generally, firms tend to consider it a market of opportunity and so it is impossible to provide a list of manufacturers. There are few United Kingdom firms, but one, William Warne & Co., has been making booms for some 25 years.

Booms are made in a variety of sizes and no one size will be adequate for coping with all spill situations. The small curtain boom with 120 mm inflated tubular flotation unit and 180 mm skirt, is ideal for coping with small spills on calm inland waters (see Fig. 5.14). There are many instances where a larger boom would be cumbersome and much more difficult to handle, particularly in confined spaces. Booms of this type are quite light, and a man can easily carry a 60 ft length.

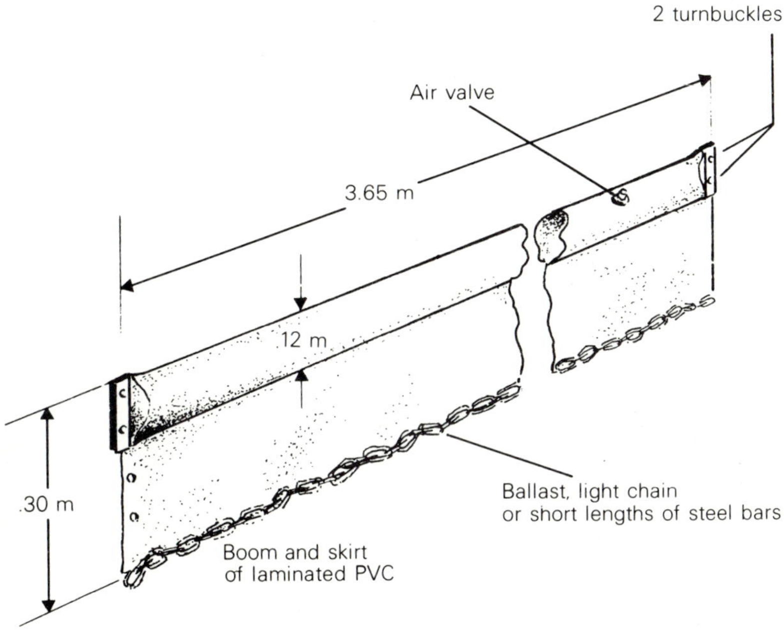

Fig. 5.14 Example of lightweight boom

For more stringent hydraulic conditions, e.g. where waves 30–45 cm in height and currents of up to 2 knots may be anticipated, a larger boom having a freeboard of 30–40 cm and draught of at least 45 cm should be selected. There are a number available in this range, and they vary in price according to the thickness of the material from which they are made depending on whether they are made for permanent or temporary use. Some of them have insufficient ballast loading and often there is no provision for modification to accommodate added ballast.

For compactness and easy storage those types using air-inflated tubular buoyancy cylinders are to be recommended. They are also less cumbersome to transport to the area where they have to be used. If possible the boom should have detachable ballast, preferably in the form of a metal chain, the size and weight of which can be selected to suit the particular site.

For permanent installation, foam filled booms are a more obvious choice as in time the outer fabric is liable to some damage either from ships or from vandalism, which might sink an air-filled boom. The larger Warne booms, (for example Fig. 5.15), although being much heavier and more difficult to transport because they are manufactured as a finished assembly and carry a heavy chain ballast, are recommended. As their installation and occasional withdrawal for maintenance is spread over long periods, the extra manpower required for deployment is no serious drawback. They are robustly constructed of heavy nylon-neoprene and have adequate attachment points for moorings along the skirt (see Fig. 5.16).

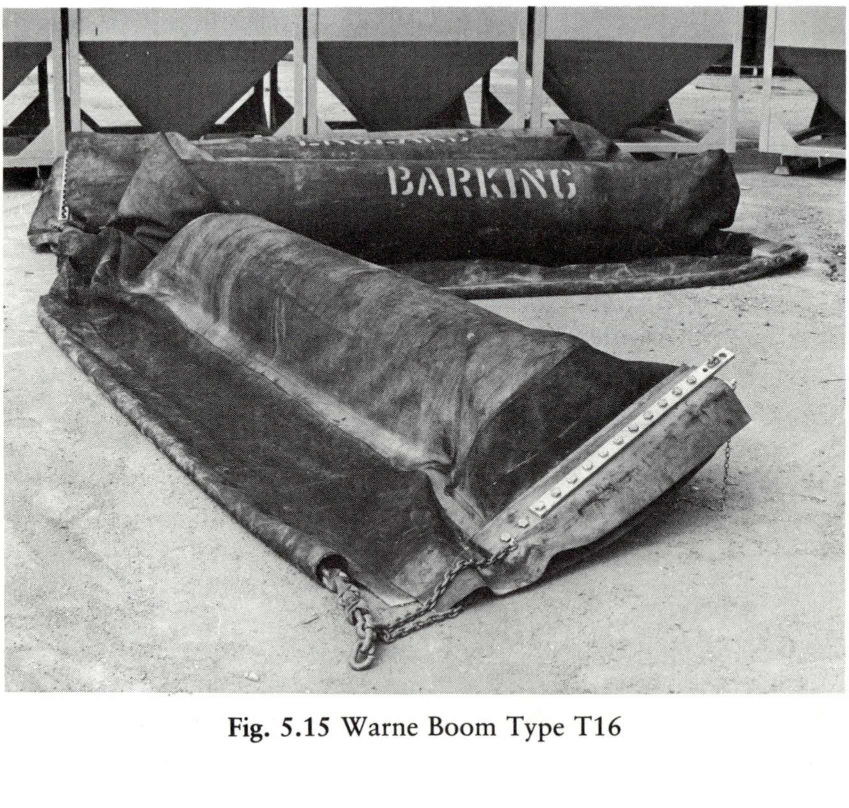

Fig. 5.15 Warne Boom Type T16

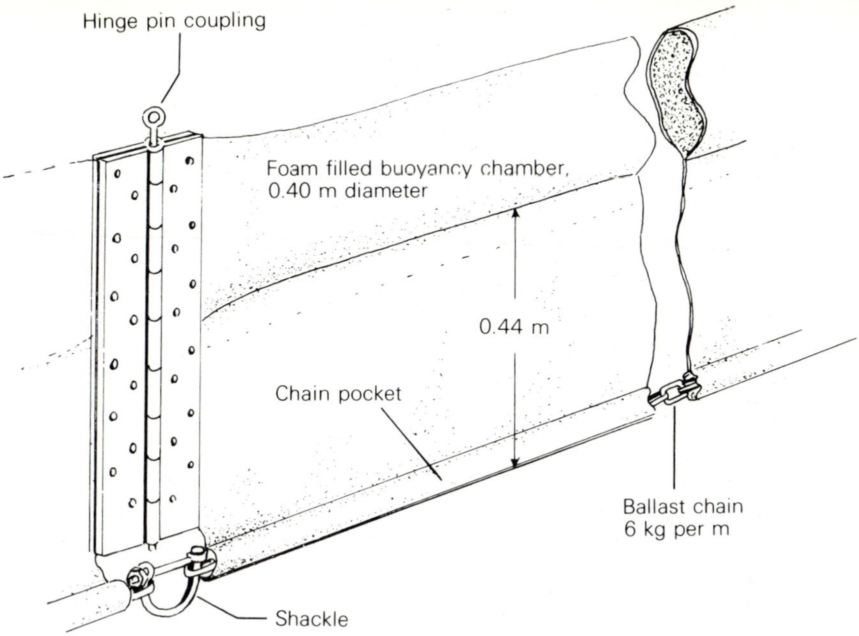

Fig. 5.16 Heavy boom for permanent deployment

In situations near oil terminals or docks where there are perhaps higher risks of pollution than in the open sea, heavier booms may be conveniently stored in zig-zag fashion on board a raft (see Fig. 5.17). The raft may be either towed to the polluted water, or if the distance is too great, the boom pulled off by a work-boat and towed to the site. This method of storage avoids the accumulation of weed growth in the boom and consequent loss of freeboard. It also avoids abrasion on piles, etc. if the boom is moored near a wall or jetty.

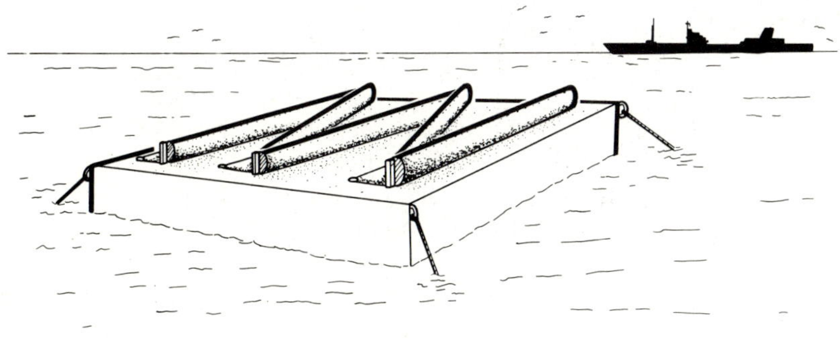

Fig. 5.17 Boom flaked on raft ready for deployment

As no one single boom design is ideal for all types of spill there may be something to be said for some sort of adjustment to be built into certain existing types. Clearly, air inflatable booms may be only partially inflated to increase their depth of skirt, and the ability to vary the ballast weights according to the situation may reduce the work in handling an otherwise heavy boom for a relatively small job in a calm area. Among the many inventions are two which are intended to deal with the disadvantage of the air inflated boom, the need for a compressor and power supply, which can be an extra load to move to a difficult site. A Swedish invention uses spring loaded frames inside the boom, which, when unclipped, expand and draw in air through non-return valves. Repacking a boom of this kind is not easy as every spring has to be compressed again. The Hydraulics Research Station developed a different version in 1973. The buoyancy tube was a lay-flat type into which was inserted a series of rigid PVC hoops. When the ends of a section are pulled apart air enters through non-return valves and the tube opens into cylinderical form. No pumped air is found necessary.

The limitations of booms

At first sight, to the novice, a boom is an immediate response to many oil pollution problems. It can be used (1) to prevent oil from spreading (2) to collect and concentrate floating oil (3) to keep oil from entering and polluting a harbour or even shore line. Undoubtedly booms have been used successfully for these purposes, but not very often. There are far more examples of oil passing under, or even over, booms, of booms being carried away by wind and waves, and of booms being completely broken up by the sea. Booms can work well but their limitations must be remembered and their placement, which is a skilled operation, must be carried out with great care.

The ability of a boom to arrest the flow of surface oil is determined mainly by the hydraulic conditions and their physical effect on the barrier. For instance, if the water currents are too strong, the submerged portion of the boom (the skirt) may be deflected upwards sufficiently to allow oil to pass underneath.

If the current is at right angles to the boom oil will have started to pass underneath when the flow has reached 0.3 m/s. This assumes deep water beneath the boom.[9] However, if the water is shallow, say only twice the depth of the submerged portion, the speed of the water passing beneath the barrier will be doubled, thus the limiting velocity in front of the boom will be 0.15 m/s (see Fig. 5.18). A small shallower, draught boom would be more effective in this case. Ideally, the depth of water beneath the booms should be at least five times its draught.

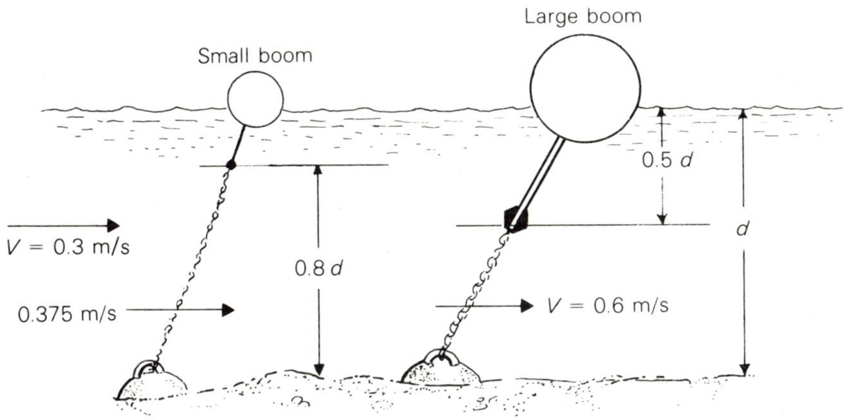

Fig. 5.18 Blockage effect of booms in shallow water

A better performance will be obtained if the boom is placed at an angle to the flow and thereby used to deflect the oil towards the side of a river or to a collecting point in an area of slower moving water. For practical purposes in keeping the total length of the boom required within reasonable limits, the finest angle between the boom and the line of flow should not be less than 14°. The formula

$$\sin \alpha = \frac{0.3}{V_{max}}$$

may be used to determine the angle the boom makes with the current direction which must not be exceeded. However, as sin 14° = 0.24 the maximum current flow must not exceed 0.3/0.24 = 1.2 m/s. As, in a river current, velocities decrease towards the banks the boom may take up the configuration shown in Fig. 5.19. Such a system can often be arranged in a tideway, but the open sea presents additional obstacles. The flood and ebb currents may not be at 180° and the sea surface itself may be exposed to winds from any direction.

Long period waves in the sea will not affect the performance of a boom provided that it is flexible, but locally generated waves (1–3s) superimposed on the main system are often steep enough to break on the boom and 'splash-over' will result. As oil is a surface pollutant much of the 'splash-over' will be oil. Clearly,

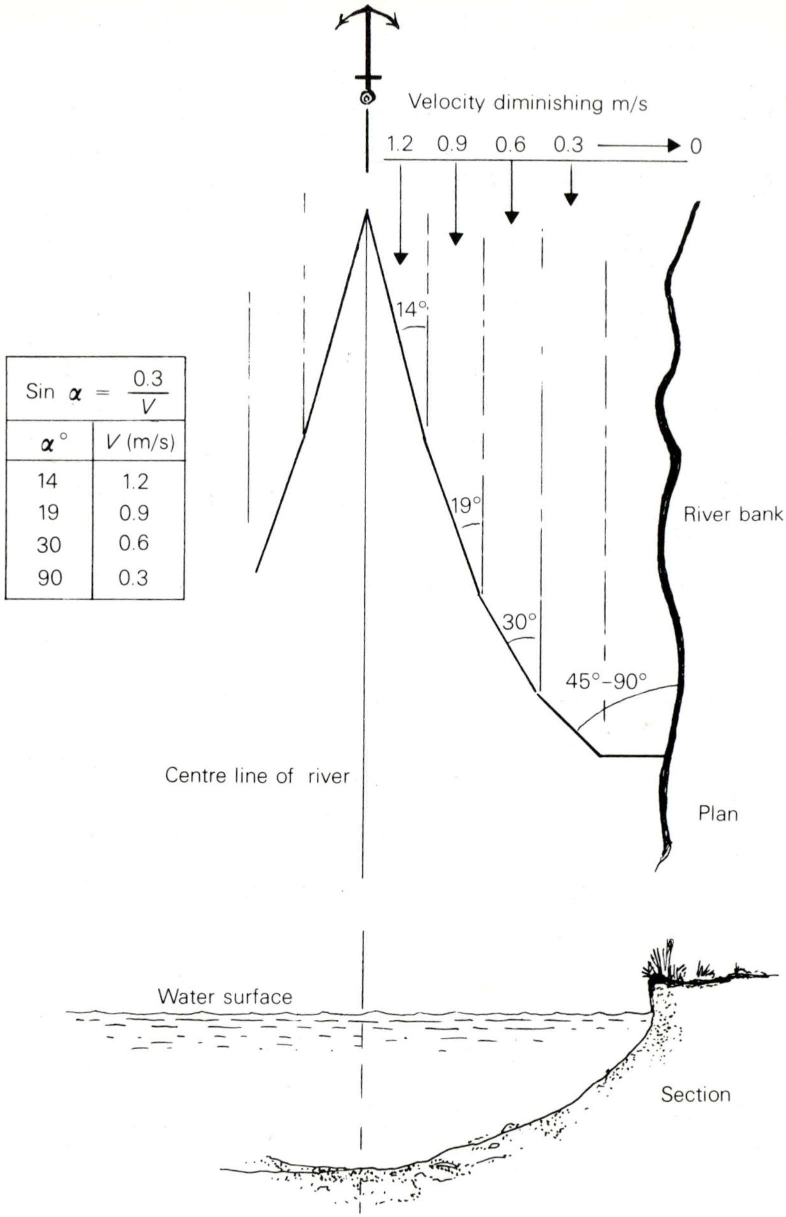

Fig. 5.19 Angle of boom limited by current

overtopping of the boom may be reduced by increasing its dimension, and in particular its freeboard. This larger boom offers more resistance to wind and waves, so the boom must be stronger and the skirt weight greater, thus requiring additional buoyancy and the whole thing becomes too big and heavy for easy and speedy deployment needing many men or special launching equipment.

In order to maintain freeboard and draught in a wave situation, a boom must be flexible. Flexibility and strength tend to be incompatible, so care is required in this aspect of boom design. Figure 5.20 shows what may happen to both draught and freeboard with a rigid boom in a sea way, and also how a flexible boom may maintain freeboard and draught. In addition wind and waves produce high tensile forces in the boom, which inhibit the free floating properties of the boom. This effect is reduced if the boom is fairly elastic, but even here the stretched boom would eventually fail if the elastic limit is exceeded. One way of getting over this problem is the Trelleborg* Universal boom. This boom has a system of load relieving lines which are somewhat shorter than the boom itself and is connected to the boom by stay-lines every four metres. This system is on the up-stream or wind-ward side of the boom greatly improving the equilibrium of the system.

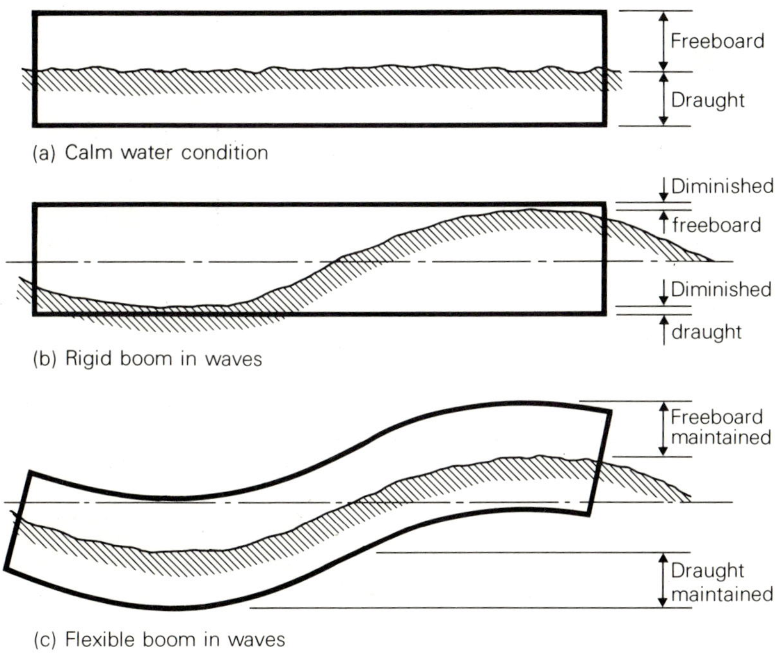

Fig. 5.20 Consideration of draught and feedboard

There are, as yet, very few records of the successful use of booms in rough waters, and although many claims have been made of the adequacy of certain types in high seas. Such claims

*Trelleborg Gummifabriks A B S 231 01 Trelleborg SWEDEN.

more usually refer to the robust nature of the construction of the boom than to its oil trapping properties. Even if containment is found possible, the removal and storage of salvaged oil is very difficult, if not impossible, in a confused sea.

It is doubtful whether the natural conditions limiting the successful use of booms, that is, in currents and waves, would permit of any increase in performance above that already established by years of research. It is known that when oil is carried by a current towards a barrier, it first collects, the thickness increases and then the oil starts to pass beneath the boom, though the oil thickness is far less than the depth of the skirt. This is because the surface currents in front of the boom will dive to avoid the obstruction and in doing so will shear off oil from the interface near the leading edge of the slick (sometimes called the 'headwave').[9] If the current velocity is small, say 0.1–0.2 m/s, a vast area of oil may accumulate in front of the boom, but as the velocity increases, possibly due to tidal action, or run-off in a river, the area is reduced and the thickness increased. The effect of the obstruction to flow caused by a floating barrier has been studied in a model tank in which particles of sawdust have been introduced into the water to illustrate the flow pattern. For the purpose of collecting oil in a tideway, the skirt should be held by the moorings in a forward position (Fig. 5.21a). If allowed to remain vertical the zone of slowly moving water immediately in front of the boom, and therefore its ability to hold oil, is reduced (Fig. 5.21b).

A freely hanging skirt will be displaced by the current and is even less effective (Fig. 5.21c).

In the open sea, the limits imposed are mainly the enormous dimensions of a boom large enough to cope with very rough conditions. But even before that stage is reached, oil floating on the surface is likely to have become emulsified, forming water-in-oil droplets, having a combined specific gravity very close to that of the water in which they float. This is even more difficult to contain and remove than crude oils.

The casual observer may not appreciate the limitations to the use of booms, but when these are fully accepted, the effective operation of booms, where conditions are suitable, has proved to be very useful as part of any system aiming at mechanical recovery of spilled oil.

In some cases, where complete recovery is neither intended nor possible, it may only be necessary to prevent oil from entering particularly vulnerable areas. In others, the concentration of floating oil by the use of booms may considerably enhance the prospect of complete recovery of oil by mechanical means (skimmers). A few typical situations will now be examined.

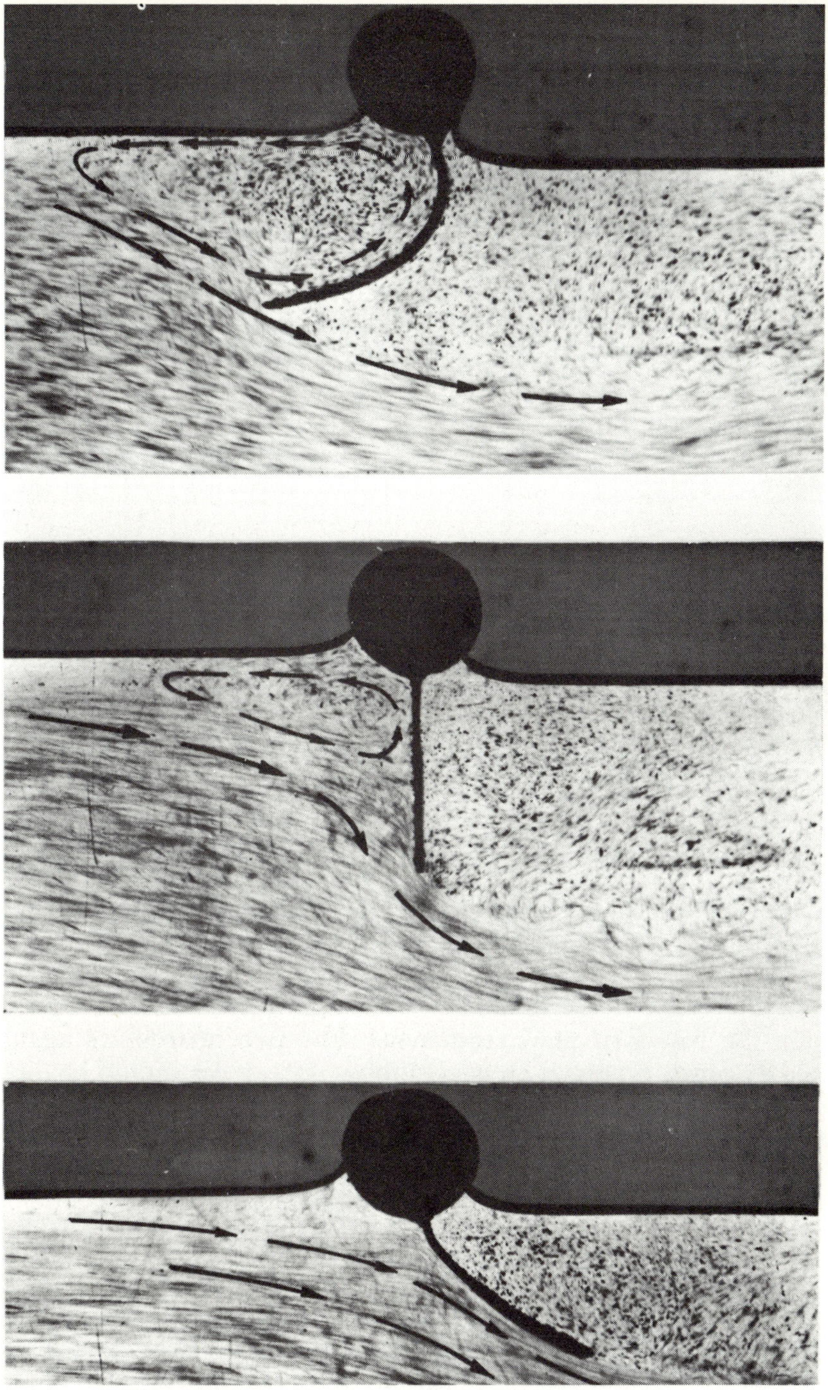

Fig. 5.21 Flow beneath the skirt of a boom

Type situations

Example 1: Oil spill into a stream

Oil is leaking into a small stream which has its outfall in a river well known for its fishing. The stream is about 4 m wide and 1 m deep at the centre. The average velocity of flow is about 0.6 m/s.

This is a case that can be dealt with efficiently by the use of simple everyday materials, provided that some method of picking-up the oil is available (a skimmer or gully-emptier).

The boom shown (Fig. 5.22) is made from 2 boards about 0.3 m wide held by a central post (C) driven into the bed at an angle (see inset) and supported by a second wooden member (D). The plan angle (G) should be about 60°. The velocity of 0.6 m/s is too high for a single boom to be placed obliquely across the stream.

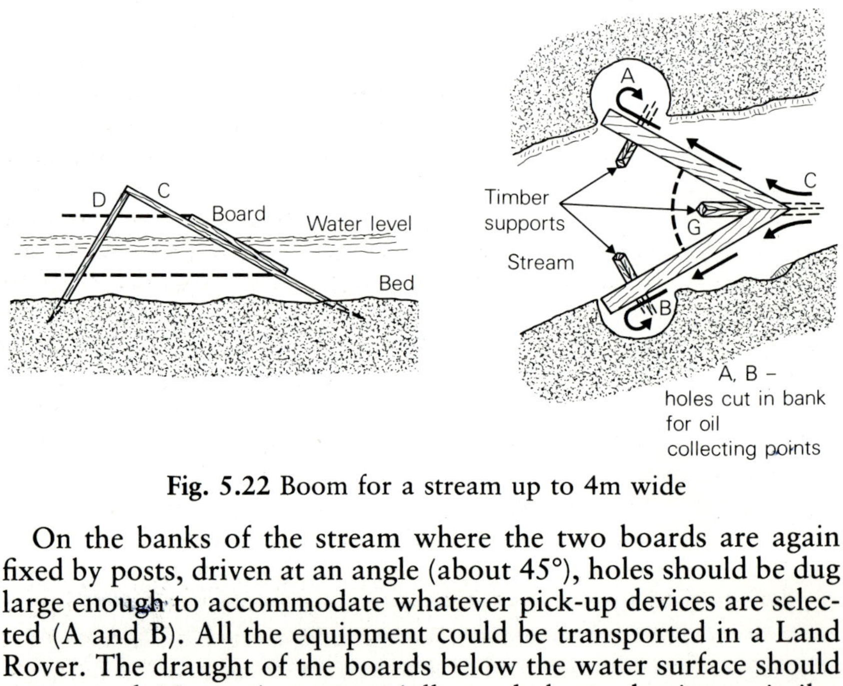

Fig. 5.22 Boom for a stream up to 4m wide

On the banks of the stream where the two boards are again fixed by posts, driven at an angle (about 45°), holes should be dug large enough to accommodate whatever pick-up devices are selected (A and B). All the equipment could be transported in a Land Rover. The draught of the boards below the water surface should not exceed 15 cm. A commercially made boom having a similar draught and moored to the centre of the stream from pegs in the bank, should be equally useful in a case of this sort. If the water level is fluctuating, a floating boom would have obvious advantages.

If in the above case, pollution is in the form of an opalescent film of oil (i.e. very minute quantities), the collecting points should have absorbent material such as straw scattered on the surface. This should be removed manually using hay forks and rakes or even fishing nets.

Clearly, from the Pollution Control Officer's point-of-view, the source of the offending leak should be established as soon as possible, but there should be no delay in deploying the necessary equipment for picking up oil already spilt.

Example 2: Oil coming up an estuary from the sea

An estuary 70 m wide and 3 m deep is threatened with pollution from offshore. The tidal range is 4 m and within a mile of the coast it has smoothly sloping banks and a mixture of sand and mud as bed material. The deepest point is at the centre and it shelves gradually towards the sides.

The strength of the currents in this estuary will tend to diminish towards the sides due to the reduction in the depth of water and the effect of the friction of the bed on water passing over it. The velocities will depend very much on the length of the river and its tidal prism, that is, the total inter-tidal volume that enters and leaves on each flood or ebb tide. The higher tides, or springs, will be accompanied by higher current velocities.

If the maximum current velocity does not exceed 1.2 m/s at the centre of the estuary, it would be possible to boom the river using two lengths of boom moored to form a 'chevron' shape.[10] The best type of boom for this purpose would be of the curtain type having a freeboard of at least 0.3 m and a draught of about 0.5 m. It would require a ballast loading of at least 5 kg per m run in order to prevent the skirt from lifting in the current. The two curves of the boom would need to be about 100 m long and moored as shown in the accompanying figure (see Fig. 5.23).

It will be seen that at the centre of the river, that is, in the region of the highest velocity, the boom is riding at a smaller angle to the flow than at its inshore ends. By this means it is able to deflect the surface flow towards the sides without losing oil beneath the skirt. As the current diminishes towards the sides the angle may become more oblique. The precise angle at which the boom should be placed in the current is defined as $\sin = 0.3/V$ where V is in m/s.

Clearly an arrangement of this type (see Fig. 5.24) will promote an accumulation of oil at the 'inshore' ends (A) and the front of the oil collected will advance back along the boom until it eventually reaches a zone where the higher velocity causes it to dive beneath the boom. For this reason the oil must be collected or

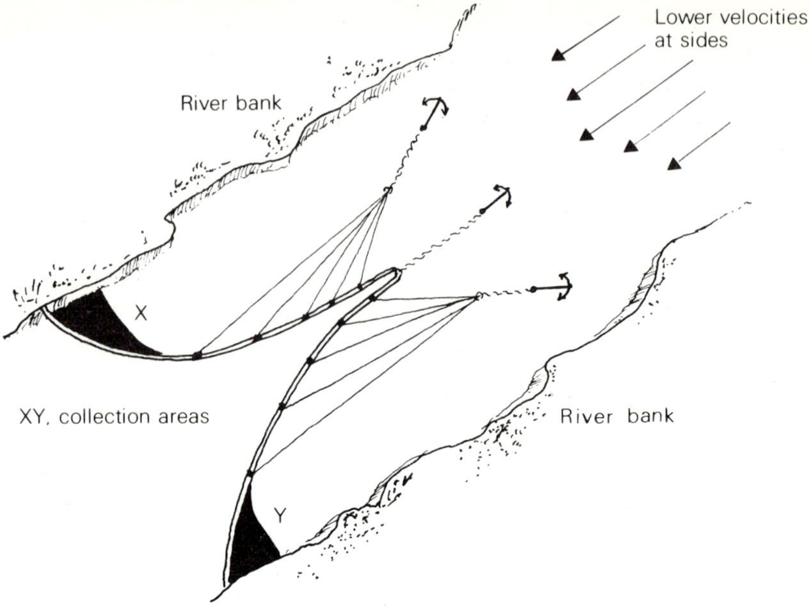

Fig. 5.23 Boom chevron for a small river

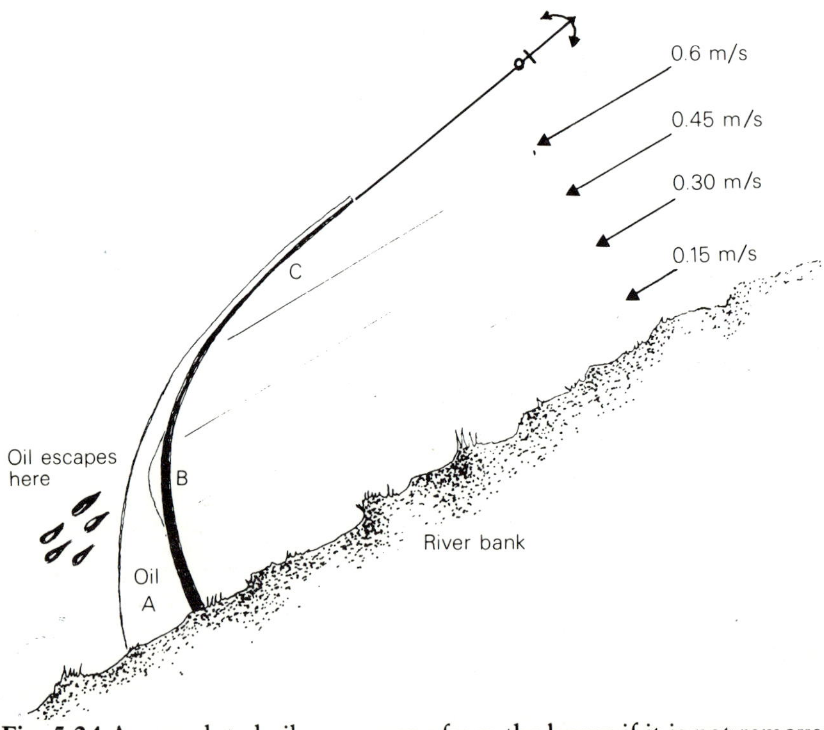

Fig. 5.24 Accumulated oil may escape from the boom if it is not removed at the rate at which it collects

disposed of before too great an accumulation occurs. The zone that ultimately limits the boom's ability to retain oil will be that where the water moves at more than 0.3 m/s (B). The thin band of oil which travels landward adjacent to the boom (C) is not affected by the higher currents, as it is in the region very close to the boom away from the diving flow. If the boom tension is inadequate then it tends to take up a J formation with the point B some distance from the bank. It has been suggested[11] that a relatively simple way of improving the shape is to attach boards to the boom so that the current drag forces the boom into a better shape. A plank 10 or 12 feet long and 12 inches deep hinged to a similar plank secured to the boom and held against the current by a chain forms a deflector (Fig. 5.25). Twelve of these can hold 200 ft of boom relatively straight in a current.

This arrangement will only succeed if the velocities slacken at the side of the channel. In deep rivers, gorges, etc. high velocities may persist to the edges and it is not then feasible to use a boom.

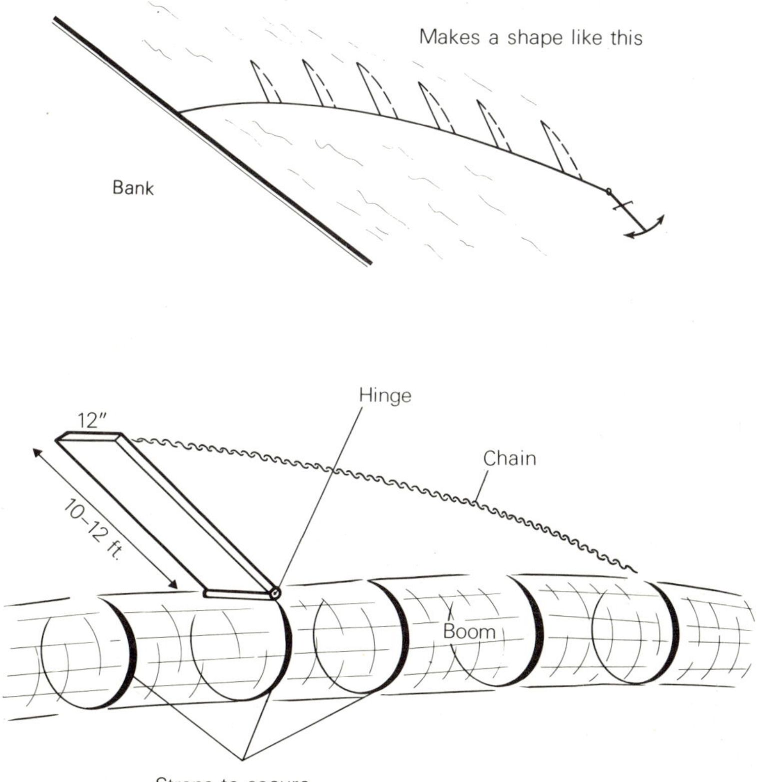

Fig. 5.25 Deflectors used to straighten the line of a boom

Example 3: protection of intakes

The use of a boom to exclude oil from fresh water intakes or intakes for power station cooling water and other similar purposes, can often be satisfactorily achieved. It is likely that there will be no requirement to collect or remove the oil only to bypass the intake. A typical example is shown in (see Fig. 5.26). The system works because, although the discharge through the intake grill may be quite high, the area over which it is spread is usually large and thus the velocities are low.

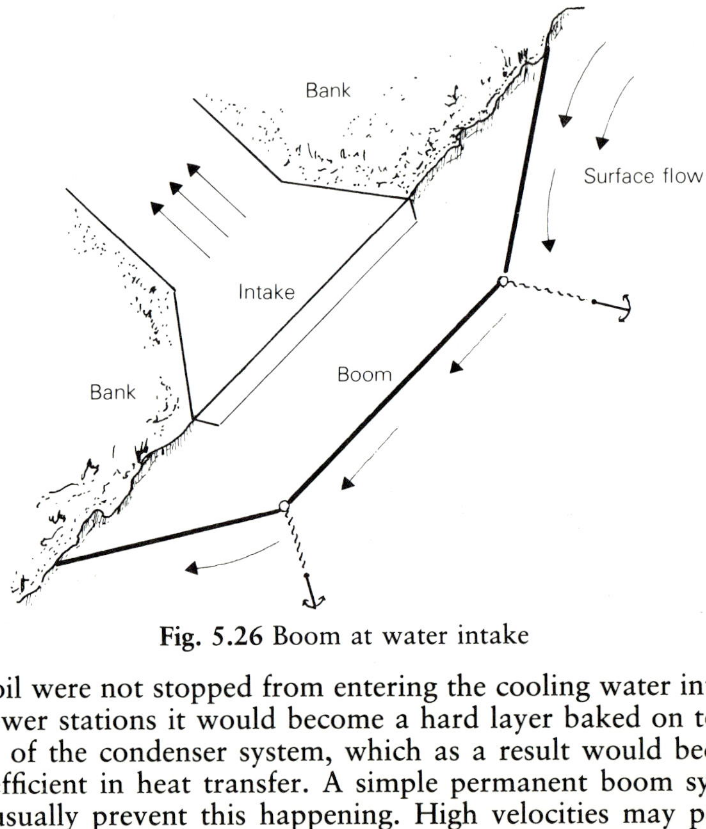

Fig. 5.26 Boom at water intake

If oil were not stopped from entering the cooling water intakes of power stations it would become a hard layer baked on to the tubes of the condenser system, which as a result would become less efficient in heat transfer. A simple permanent boom system can usually prevent this happening. High velocities may persist just in front of the intake screen and the boom should be placed a few metres away where the cross-sectional area is greater and the velocities smaller.

In some areas the cooling water intakes are exposed to long fetches of water and are subjected to considerable wave action. Clearly, a feasibility study of individual sites must be made. The velocity distribution across the inlet should be measured at low water when the cross-sectional area is least, and thus the intake velocities highest. If possible the boom should be placed some distance away from the intake structure to avoid damage through chafing contact with the piles or screen. The moorings should be of adequate length to allow for tidal rise and fall. A typical arrangement is shown in Fig. 5.27.

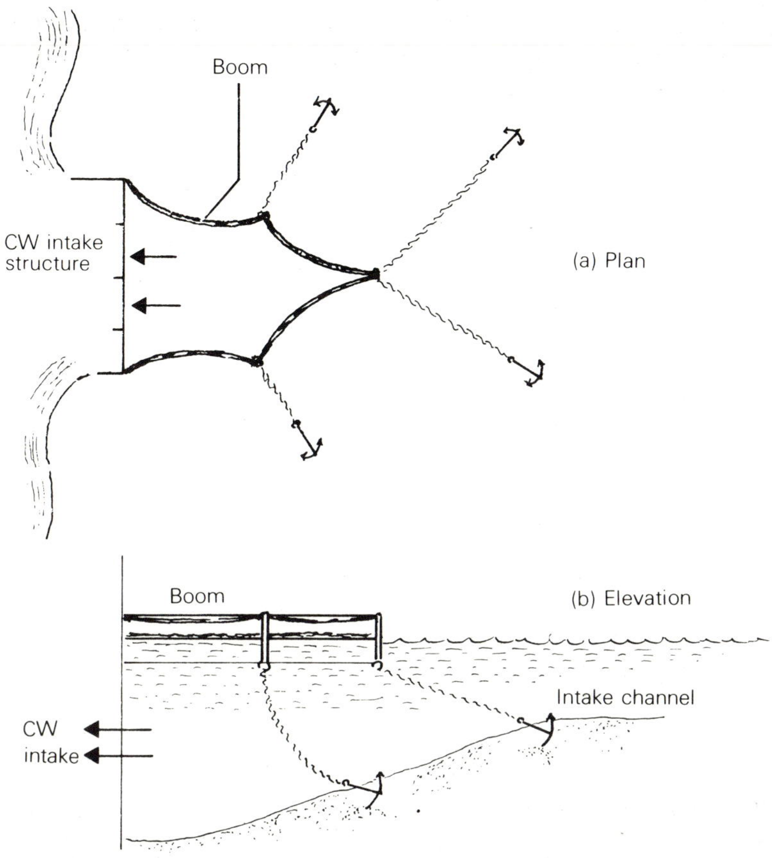

Fig. 5.27 Boom at a cooling water intake

In at least one instance, where a boom has been placed across a cooling water intake, the author has observed that lack of maintenance has resulted in considerable weed growth on the boom causing an unacceptable loss of freeboard. The rate of marine growth will vary from place to place and according to the time of year and water temperature. To operate satisfactorily, regular inspections must be made, as a loss of freeboard of only one section will result in the whole scheme becoming inefficient.

Example 4: Boom used to corral or collect oil in open water

To encircle or control oil slicks on open water a boom of the Vikoma type is very suitable. In order to control the system, to the required degree, a work boat is required at each end of the boom and the crews need training in 'station keeping' in order to hold the boom on the best heading in relation to the oil slick.

It is possible to use an oil pick-up device within the collected pool of oil. The Vikoma Seaskimmer (see p. 131) was developed for this purpose. An alternative is to leave a gap at the apex of the V and fill it with a suitable skimmer, either in, or just aft of the gap.

Clearly, the contraction formed by the 'V' or 'U' will tend to increase surface velocities locally and the speed at which the system is maintained relative to the ambient flow conditions will have to be carefully controlled. In rough weather it may be difficult to reach the high standard of co-ordination required between the various craft involved.

The towed 'V'-boom system would not necessarily employ Vikoma booms. Any curtain type boom large enough to withstand overtopping by waves would be suitable for this type operation.

It might be difficult to operate the usual weir-type skimmer in this type of exercise owing to the difficulty of keeping the weir at the correct level, the 'weir boom' described earlier may be effective otherwise absorbents or disc-type devices might be more suitable (see Fig. 5.28).

If the boats move too fast against the current, some oil will escape beneath the boom and most of it will miss the skimmer. The relative velocity between the tug-boats and the opposing current should not exceed 0.5 knot. If the current is too strong, the system can still operate by allowing the boats to drift sternwards and maintaining the same differential velocity. If possible the entire system, having collected oil, should be moved to a sheltered area, perhaps in the lee of a headland or a large tanker, where the oil may be removed. If the spill is large or continuous, more than one team and boom assembly will be required.

It will be seen (see Fig. 5.28) that the two sides of the boom chevron take up a curved shape in the 'towing' position. If the drag forces are high enough the assembly may form into a semicircular shape and the boom will then be presenting some of its length at 90° to the current or direction of travel. Under these circumstances, oil is very easily lost beneath the boom. A much improved performance can be achieved by adding thin wire bridles at regular intervals so that the boom maintains a 'V'-shape under tow (see Fig. 5.29).

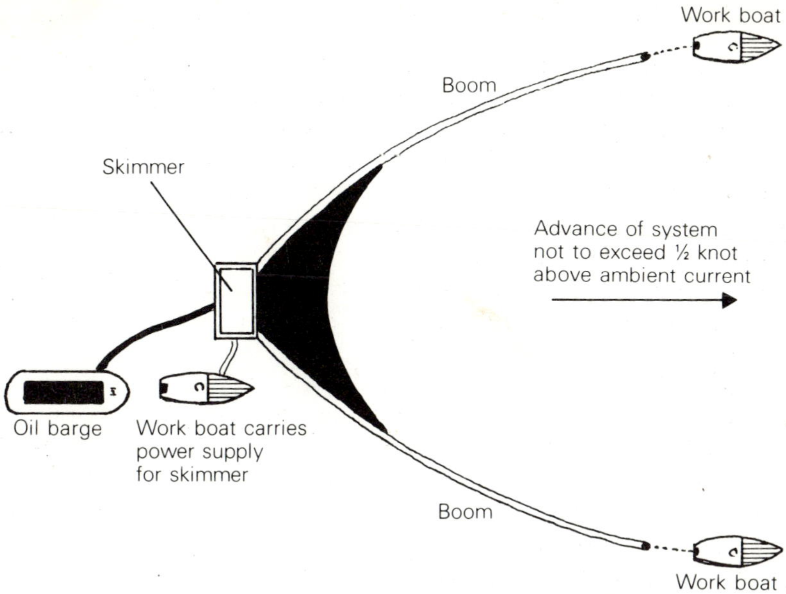

Fig. 5.28 'V' boom coralling and skimming

The oil pollution team and boom deployment

There have been many oil spills, the more spectacular having been well documented. There have also been many lessons learnt and some of these too have been recorded. But it is unfortunate that in spite of the experience gained from the variety of incidents that have occurred the same mistakes are made whenever a spillage occurs at a new location.

It may be due to the random nature of accidents that many centres which have been the victims of pollution in recent years have had little or no equipment with which to deal with it. The occurrence or recurrence of pollution in a single location can usually be reckoned as unlikely, but all the time that liability, however remote, exists, some provision should be made to deal with it.

Chemical treatment may have much to offer: it is easy to transport and uses light equipment, so it can be available at a site within a few hours.

On the other hand, oil booms may take up a considerable amount of storage space, and be cumbersome to handle and transport. Only in the case of small spills can improvised booms be made up and deployed quickly.

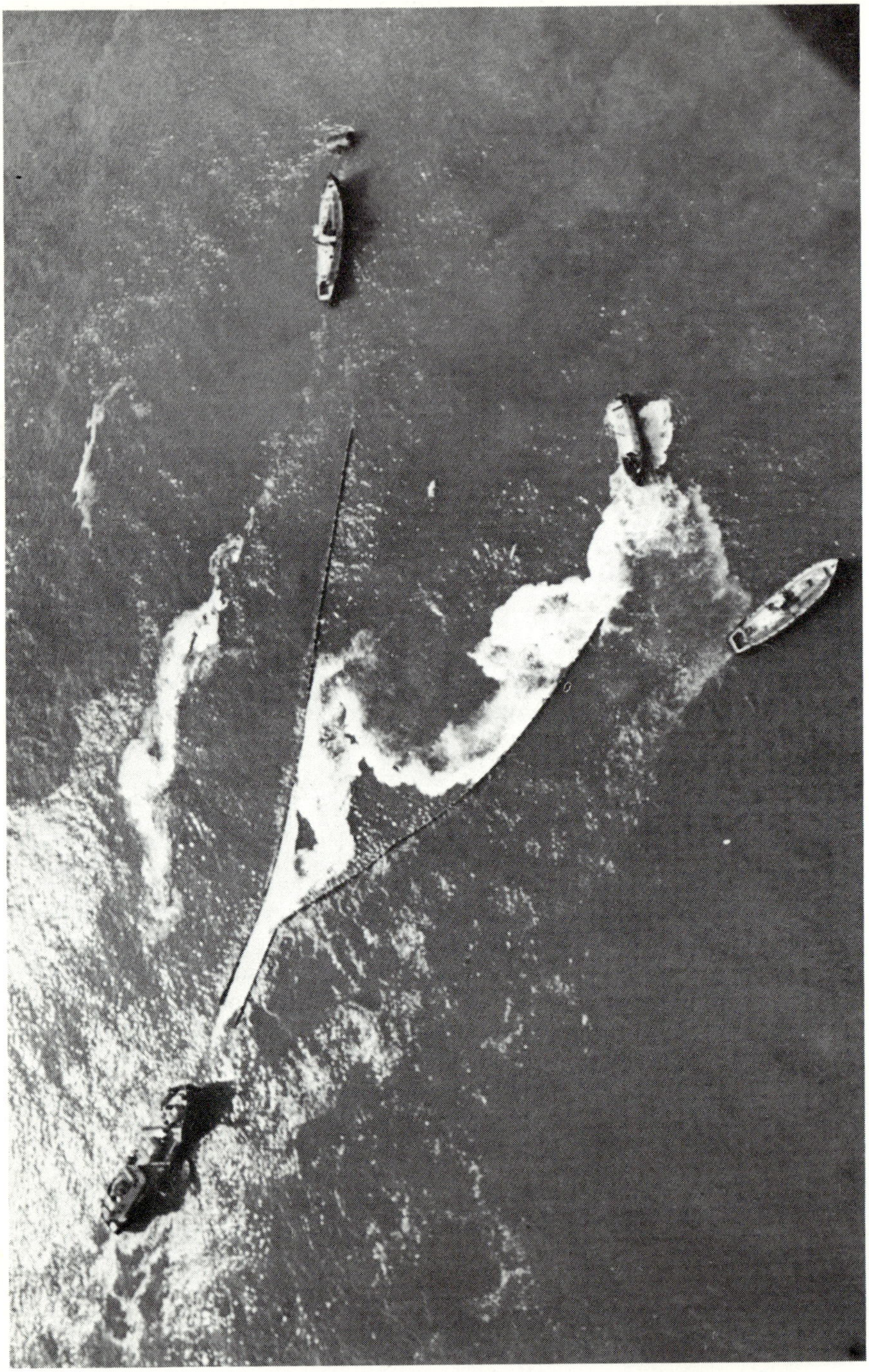

Fig. 5.29 'V' type boom with bridles

The answer lies in having suitable booms available at storage points situated within easy reach of ports and terminals which have high liability to pollution. It may not be possible in all such areas to operate booms owing to the local hydraulic conditions, but where booms can be used they should be made available—physical removal of oil is preferable to chemical methods of dispersal.

It is important that teams of men to operate craft and booms have plenty of practice in handling the equipment and that they appreciate the mechanical and physical limits of the items they are handling. Particularly vulnerable sites should be surveyed and if protection by booming is feasible then a trial should be undertaken and a careful note made of the exact position of anchors. In some cases ring bolts can be fitted to convenient rocks at the shore end. If the boom is needed in a real oil spill then it can be positioned in the same, correct place, without delay.

References

1. NEWMAN, D. E. and MACBETH, N. I., 'The use of booms as barriers to oil pollution in tidal estuaries and sheltered waters', IP/WHO, Seminar on Water Pollution by Oil, Aviemore, 1970.
2. MAINVILLE, C. R., *State of the art review of oil containment for use at offshore terminals*, Ocean Science & Eng.Inc., Rockville, Maryland, November 1973.
3. BULSON, P. S., *Currents produced by an Air Curtain in deep water.* The Dock and Harbour Authority, May 1961, pp. 15–22.
4. HOPPER, W. C. and MARCINOWSKI, H. J., *Removal of oil from water surfaces*, Stichting Concawe, The Hague, 1969.
5. BARGER, W. R., 'Laboratory and field testing of surface film-forming chemicals for use as oil collecting agents', Proc. Joint Conf. on prevention of oil spills, Washington DC, March 1973.
6. GARRET, W. D., 'Confinement and control of oil pollution on water with monomolecular surface films', Proc. Joint Conf. on prevention of oil spills, New York, December 1969.
7. RIJKSWATERSTAAT, Shell Herder Trials, Hague, July 1974.
8. WILSON, H. B., 'Development and testing of a weir boom for Oil Recovery at Sea', 1981 Oil Spill Conference, Atlanta, Georgia, USA.
9. WICKS, M., 'Fluid dynamics of floating oil containment by mechanical barriers in the presence of water currents', Joint Conf (API/FWPCA) on the prevention and control of oil spills, Washington, 1969.
10. NEWMAN, D. E. and MACBETH, N. I., *Oil booms at a tidal inlet*, HRS Report No. INT 118, October 1973.
11. GREEN, G. D. *et al.*, Use of deflection for deployment of oil booms in rivers Oil Spill Conference, 1979, p. 41.

6

Mechanical Devices for the Collection of Oil

Collecting oil, as distinct from sinking or dispersing it, has generally been seen as the most desirable way of dealing with spilt oil, at least in theory. In this chapter consideration is given first to those devices that have been proposed or described for the removal of oil from water. They are known collectively as 'Skimmers', and form a central part of an overall system for recovering oil from the surface of rivers, canals, lakes, harbours or open water. The latter part of the chapter is devoted to a consideration of those factors that are particularly relevant to the collection of oil from the open seas, as distinct from enclosed or quiescent water.

Several hundred mechanical devices for the removal of oil from water surfaces have been proposed, and many have been described in detail, although scarcely more than a dozen have come into common regular use for oil removal, or indeed even reached the production stage.

Although considerable ingenuity has been shown in the mechanical design and construction of these devices, the principles upon which the oil pick-up capability are based are relatively few. These exploit the differences in physical behaviour that exist between oil and water—mainly differences in gravity, adhesion to a solid surface and absorption into a porous material. These differences, and combinations of them, have given rise to the basic designs of skimmers for use on rivers, in enclosed waters, harbours, estuaries and open seas. The basic principles, and examples of the skimmers produced from them, are described below.

Knowledge of the performance that can be achieved in practice by these skimmers is frequently difficult to obtain, and all too often the design data or tank testing results are quoted as 'obtainable performance'. Experience of skimmers at sea under realistic conditions has shown that many of these devices do not, and probably cannot, achieve the results claimed on the basis of tank testing performance. In some instances this is clearly due to a failure by the designer to build for the conditions that are encountered in practice. Some of the factors that must be considered are discussed below.

It must be realised from the onset, that the skimming of oil takes time to organise, and to carry out; during this time, the oil does not remain stationary—it moves and it spreads, and it also absorbs water. Where a natural river-flow exists it can be used to carry the oil to an area where, by virtue of the slower currents, it can accumulate to a reasonable thickness in a coherent layer prior to skimming. But since the extent to which the movement and spreading of oil at sea can be controlled or prevented is very limited, it follows that skimming operations at sea must be conducted in such a way as to be consistent with oil movement, and to take account of the spreading that inevitably occurs. Both the movement and spreading of oil are affected by the incorporation of water with the oil. This may proceed rapidly, and in some instances the character of the discharged oil will change markedly within hours.

Even with the most effective boom system to confine and concentrate the oil layer, the thickness seldom exceeds a few inches. At this thickness, any high capacity skimmer will rapidly clear the oil from an area around it. The performance of the skimmer will then be determined, not by its own pick-up capability, but by the rate at which new oil can be brought to the skimmer, or how rapidly the skimmer can be moved to successive new patches of oil.

The optimum performance of skimmers is related to the properties of the oil being collected, and particularly to its viscosity which changes rapidly as water pick-up occurs. Skimmers relying on the use of absorptive belts are more effective with lighter oils, while the performance of disc skimmers deteriorates with both light and heavy oils, being at its best in the intermediate range. Other factors affecting the performance of the skimmer include the extent to which the oil progressively forms a water-in-oil emulsion, and the extent to which the oil is contaminated with dispersants or with debris.

In considering the behaviour of skimming devices under the full range of sea conditions that might be encountered, an important parameter is wave height. The tolerance to wave height is greater with some devices (disc skimmers for example), than with others (weir and centrifugal skimmers). A skimmer of any kind is likely to be ineffective if waves break over it. This point is discussed again with particular reference to the Warren Spring designed skimmer (p. 148).

In the open sea, the swell is also important. A long swell, even of considerable amplitude, is likely to have little effect upon a short device which will remain at the surface and only tilt. A skimmer that has a dimension greater than the swell wave length will not maintain its position relative to the sea surface.

Because general purpose skimmers have so far not been produced, most of those currently available have been designed to operate in one particular situation. The number of design characteristics that can be varied is small—principally the overall size of the pick-up unit (for example, the length of a weir or the diameter of the discs in a disc skimmer, which tend to dictate the overall size of the complete skimmer) and the number of such pick-up units incorporated (e.g. the number of weirs or discs). Together these define the oil pick-up capacity in relation to any particular oil.

Types of skimmer

Although a considerable degree of ingenuity has been shown in inventing devices for the recovery of oil from water surfaces, most of those produced rely upon relatively few principles. For convenience, it is possible to divide these into groups, although there is a certain amount of overlap between them.

Group I. Skimmers that rely on the adhesion of oil to a solid surface, as, for example, discs of metal or other material that are rotated in a vertical plane with part of the lower half of the disc below the surface of the water. Oil adheres to the disc and can run or be scraped off as it is carried above the surface of the water, and then subsequently collected.

Group II. Skimmers that rely on adhesion to a continuous flexible belt drawn through the oil/water interface. Oil adhering to the belt can then be scraped from it for collection. This group also includes skimmers that use continuous belts of absorbent material from which the oil can subsequently be recovered by squeezing.

Group III. Centrifugal devices that rely upon increasing the effect of gravity by creating a vortex to increase the thickness of the oil layer at a point from which it can be pumped.

Group IV. Devices in which the oil is separated from the water by passage over a weir. This group includes devices with weirs ranging in length from a few centimetres to up to 10 m, devices with multiple weirs, and devices with a variety of arrangements intended to keep the weir aligned with the sea surface.

Group V. An omnibus group including devices based upon other principles or a combination of one or more of the other devices.

Group I: Adhesive surface devices

Devices in this group rely upon the adhesion of oil to a solid surface. Almost any material with a hydrophobic surface can be used including plastic materials. The simplest device is probably a rotating drum fitted with a scraper blade to enable the oil to be collected. Such a device is, however, far from simple mechanically and has a poor oil recovery capacity, as the surface area available for oil adhesion is low.

A considerable improvement in oil pick-up capacity can be obtained by using a series of discs in place of the drum. Successful commercial skimmers have been produced with discs along a common axis, in the form of a toroid and in other geometric arrangements. Scraper blades are required on each disc, together with some mechanism for removing the oil that is scraped from them.

As with all other devices that rely upon the adhesion of oil to a solid moving surface, a certain amount of water will be picked up by the discs and subsequently be collected with the oil. In general terms, the more rapidly the discs are rotated, the greater the oil pick-up and also the greater the amount of water collected. However, in practice, a maximum value is soon reached for oil pick-up rate, this being determined by the rapidity with which new oil reaches the disc to replace that which has been removed by collection. On at least one device, oils of high viscosity tend to build up on the grill exposed to the surface, rather than flow into the openings provided. This is especially so with wet, heavy oils; these have been observed to coat the outside of the drum, and then be flung off as it rotates, rather than be carried towards the collecting scrapers.

As the viscosity increases, oil does not run from the adhesive surface so rapidly; the collecting rate for heavier oils is therefore greater than for thin, more fluid oils. However the rate of replenishment at the sea surface is correspondingly slower as the viscosity rises. The overall recovery from a series of test runs in the laboratory tank is shown (Fig. 6.1) in respect of a Vikoma skimmer.

It has been postulated that there is a speed of rotation, some-

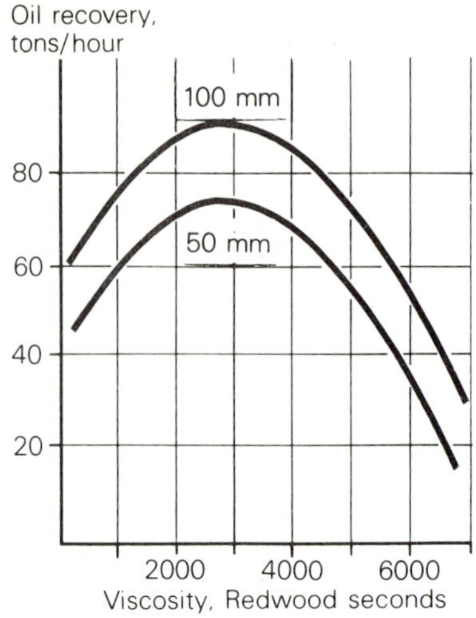

Fig. 6.1 Oil recovery in relation to viscosity

what below that which gives maximum oil recovery, which gives a maximum oil/water ratio. This is difficult to observe in practice, as in the tank tests water-in-oil emulsions tend to form, particularly as the oil is recycled, tending to obscure the free water pick-up rate, as well as markedly increasing the viscosity of the oil.

For general use, skimmers are required to pick up not only oil, but oil which contains water in this emulsified form; the skimmers currently in production certainly do this, right up to the fully developed 'chocolate mousse' emulsion which may contain as much as 80% water. It should be remembered however that these emulsions exhibit marked non-Newtonian behaviour, and they will not flow towards the skimmer discs as readily as crude oil, nor pump so easily.

In summary, therefore, these devices can be used for the recovery of oils and emulsions of a wide range of viscosities. They do however give greatest recovery rate with oils in the middle range, i.e. of moderate but not excessive viscosity. They are at their best when deployed in a pool of oil of a reasonable thickness which is continually being replaced as it is removed by the skimmer. The effect of oil layer thickness on pick-up rate is shown in Fig. 6.2.

A number of prototype devices have been constructed based upon the use of rotating drums, but few of these were of sufficient oil-removing capacity to be interesting. Progressive development of disc devices has shown that high-capacity skimmers can be produced using this principle. These include the following.

Clean Sweep (Lockheed)[1] has multiple, aluminium alloy discs

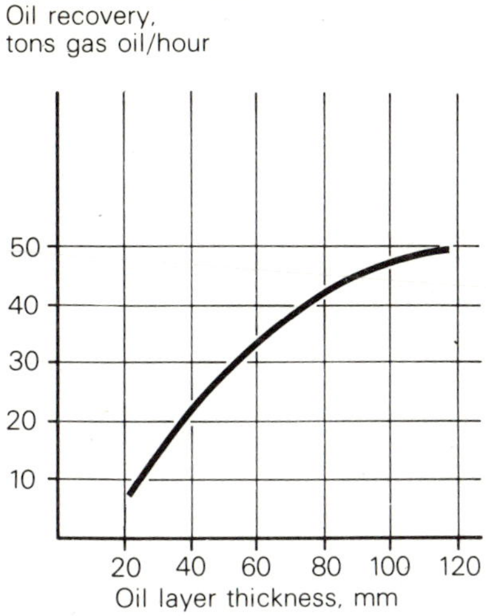

Fig. 6.2 Oil recovery in relation to longer thickness

arranged longitudinally on a common axis; the oil is scraped from the metal surfaces and flows to a central trough from which it is removed by a screw conveyor.

The Clean Sweep is undirectional, in that oil is removed from the sea surface on only one side of the device, and any oil that is not collected by the skimmer, will, by the rotation of the barrel, be driven underneath and thus be lost. The rotation of the barrel itself causes a 'paddle-wheel' circulation of sea water and effectively drives the device forward at slow speed.

Vikoma Skimmer[2] has discs of either a plastic material or aluminium alloy, approximately 1 m in diameter, arranged either in the form of a toroid or in four banks forming a square (Fig. 6.3). The discs are driven by a hydraulic motor, powered from a separate diesel-driven hydraulic generator connected to the skimmer by an umbilical containing the flow and return hydraulic fluid pipes. Scraper blades allow the oil to drain down oil ducts, from which it flows into an oil bowl, and thence via a hydraulic oil delivery pump via another pipe in the umbilical, to a storage tank or temporary oil container. The level of the discs in the water is set by a level sensing device which controls the water level in a central buoyancy chamber.

In operation, the skimmer floats in the middle of a pool of relatively thick (several inches at least) oil, when it collects the oil from all directions. The recovery rate of the larger model (Vikoma 100) with oils of medium viscosity is 100 tons per hour. As with all disc skimmers, a certain amount of water movement is induced below the skimmer. The largest Vikoma skimmer has been de-

Fig. 6.3 Vikoma skimmer, with crane transported on vessel

signed for high seas use, and is not suitable for use in shallow waters as it requires a minimum draft of approximately two metres. It has a power unit which can be mounted together with a hydraulically operated crane on a base plate. The whole unit is suitable for attachment to the deck of a small tanker or even a tug, so that the skimmer can be raised from the transporting vessel and lowered into the sea, within a boom if required (Fig. 6.4). The Vikoma 50 requires a draft of less than 1 m for effective operation.

For operating in shallower and more sheltered waters the somewhat similar but smaller Komara (Figs 6.5 and 6.6) skimmer can be used. This is designed for dealing with small oil spills under less severe conditions, such as those pertaining to oil ports, harbours, estuaries and inland waters. It is a lightweight device, weighing only 100 lb, so that it can readily be handled by two men. It operates in a draught of only eight inches. With 32 discs, each 280 mm in diameter its oil recovery rate with oils of medium viscosity is 10 tons per hour. It is operated by compressed air which can be provided by compressors such as those readily available.

Care must be taken in the cleaning of all recovery devices that

Fig. 6.4 Vikoma skimmer deployed within a boom

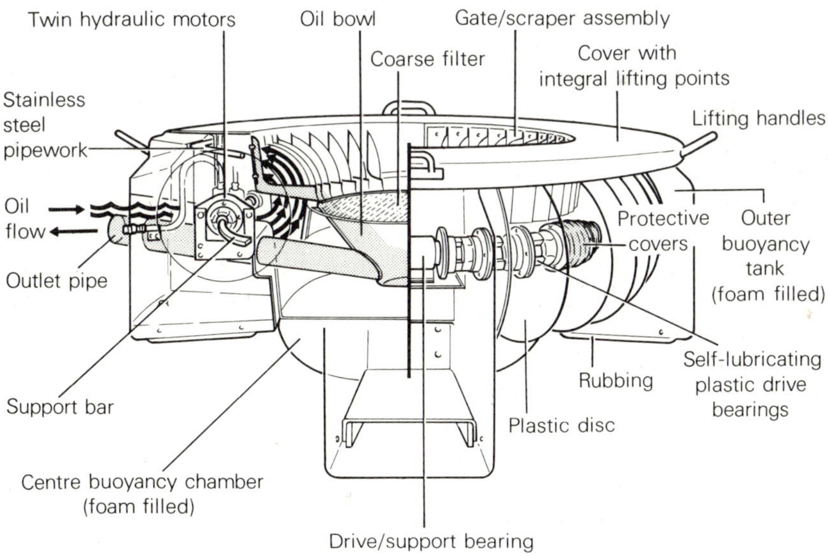

Fig. 6.5 Komara skimmer and pump

Fig. 6.6 Komara skimmer floating

rely upon the adhesion of oil to a solid surface. This adhesion is easily destroyed by detergents which should therefore not be used to clean oil from the discs. Nor should such devices be used in conjunction with oil spill control operations in which dispersant is sprayed. When cleaning is required, paraffin (kerosene) should be used.

It is clear that trash of all kinds, beer cans, bottles, dunnage, netting, plastic bags, seaweed and paper could enter the skimming devices such as those discussed above, and could give rise to mechanical damage to the discs themselves or to their driving mechanism. Experience has shown that even quite large pieces of wood do not seriously affect the performance of disc skimmers and that if damage is done to one or more of the discs, this does not usually prevent the skimmer from working nor seriously lower the skimmer recovery performance. More important is the tendency for trash to build up at the water surface close to the skimmer and prevent or hinder the flow of oil across the surface to the skimmer.

Group II: Continuous belts

This type of skimmer, consisting of a continuous belt deployed between the sea surface and an oil collection point, can be regarded as a conveyor, physically lifting the oil and allowing the adhering sea water to drain while transporting the oil to the collector. For some of the devices this is undoubtedly true. The highly viscous oil encountered during the *Arrow* spill, off the coast of Nova Scotia, mixed with debris and present in large discrete lumps, could certainly be collected in this way. For use with oils of this kind, the material of which the belt is made is not of great importance, provided it has the ability to deal with trash.

However, as the viscosity of the oil decreases, the necessity arises for greater adhesion to the surface of the belt. This introduces the need for a scraper to remove the oil. With light oils of low viscosity, even adhesion is insufficient and the belt requires absorptive properties. The scraper must then be replaced with a device such as an additional roller, to squeeze out the oil.

The simplest skimmer of this kind is a drum covered with absorbent foam. It can be used for very low viscosity oils, but the recovery rate is low, and most foam materials with the required absorptive properties, suffer mechanical damage on repeated passage through the squeeze roller, particularly from that inflicted by trash entrained in the oil. Belt materials used or proposed include stainless steel, nylon, polypropylene, other synthetic materials, felt, natural fibre of various kinds and paper. Various configurations have been suggested including thick and thin, broad and narrow belts, some with bristles, ropes and tubes.

One of the most interesting was developed by Martin Marietta[3] and marketed by Marco. Figure 6.7 shows the MARCO Class V which is a 36-foot (11 metre) skimmer equally effective for open ocean or harbour oil spills. It can be transported to an embarkation point and then towed at high speed to the spill. A 360° rotatable thruster provides both propulsion and steering, enabling the skimmer to spin on its axis. This is important for manoeuvring in harbours and from one spill patch to another. A 100 hp Diesel provides propulsion and hydraulic power.

Oil recovery and off-loading operations can be accomplished simultaneously at a rate of up to 300 gpm. Under calm conditions its performance is good and provided that sufficient oil is encountered, the design oil recovery rate can be achieved up to the viscosity limit for adhesion. Beyond this, solid and semi-solid oil will be conveyed without adhesion to the belt.

The belt material is a porous oleophilic foam backed by a tough nylon and bronze web that picks up everything in its path: distillates, bunkers, weathered crude, sorbent, and floating debris. Oil and debris are conveyed to a squeeze roller and scraper where the debris is separated and oil falls into the sump.

Fig. 6.7 Marco skimmer

A submerged induction pump behind the porous belt assists in drawing oil and debris to the skimmer when the vessel is in a stationary mode. During free skimming, the induction pump eliminates head wave problems, preventing oil from being pushed away from the belt.

Waves pass through the porous belt as oil is entrained; waves are not 'reflected' as occurs with solid belt systems. Similar adhesion problems were noted with the Sandvik Piranha Oil Skimmer which uses two perforated steel belts enabling heavy oils to be collected—water and light oils escaping through the perforations.

An alternative method designed primarily, but not exclusively, for calm waters uses a long belt or rope of absorbent oleophilic material which floats on the surface of the water. The ends are joined to make a continuous loop and this loop is brought on to a suitable vessel or the bank of the river and there passed through rollers to expel the oil and any entrained water. The remote end of the belt is usually held in position by a simple pulley, either anchored or secured to a boat. This sort of device works best with light and medium grade oils. Heavy oils coat the surface of the rope or belt and the pick-up is very much reduced. Although the principle is simple, difficulties arise in making a choice of material for the belt or rope is far from easy. Most materials rapidly lose elasticity in the repeated squeezing and the oil recovery is then much reduced. However, some materials with a reasonable life and pick-up capability have been specially developed.

A device of this kind, known as Oil Mop (Fig. 6.8),[4] is available in sizes characterised by the size of the continuous absorption band. The water content of the recovered oil was found to depend upon the opportunity afforded for the free water pick-up by the rope to run back as the rope is pulled out of the sea and on board the vessel. Some debris is collected by the Oil Mop, but there is little damage to the device itself, except by abrasion of the rope belt. Oil Mops are available from Oil Mop Inc. (and its subsidiaries) and Oil Recovery International. Attempts to use endless-loop rope-mop systems on the high seas, involving paravanes, have not been entirely satisfactory, and the idea now appears to have been dropped. The Force Seven system of Oil Recovery International, which incorporates a paravane, is designed around the casting and subsequent recovery of a number of lengths of rope mop. These lengths are separated from one another and all maintained in a lateral position with respect to the deployment ship. The performance of this system under the adverse conditions frequently existing at a massive oil spill at sea is not known, but the system clearly does not suffer from the inertial mass and rigidity problems to which many other systems are subject. The amounts of oil recovered are likely to be small where the oil layers are of the order of 0.1–1 mm in thickness, but larger amounts may be collected per cast in thick oil layers.

Fig. 6.8 Oil Mop in use

Another skimmer designed for calm water and inland waterways is the ZRV (Zero-Relative-Velocity) Oil Skimmer[5] which consists of a flat (Fig. 6.9) endless absorbent belt of composite composition (Astroturf and polypropylene), positioned to float like a blanket on the water surface. In theory, since there is no appreciable velocity difference between an oil layer on the water and the flat absorbent belt, there will be no major loss of oil, as the belt will conform with surface waves. Performance data for the oil recovery of the prototype ZRV skimmer under oil spill conditions is awaited.

Group III: Centrifugal devices

Historically, the first commercial separation of oil and water was probably the skimming of cream from milk, for which centrifugal separators were an early development. It is therefore not surprising that a number of inventors have tried to use various modifi-

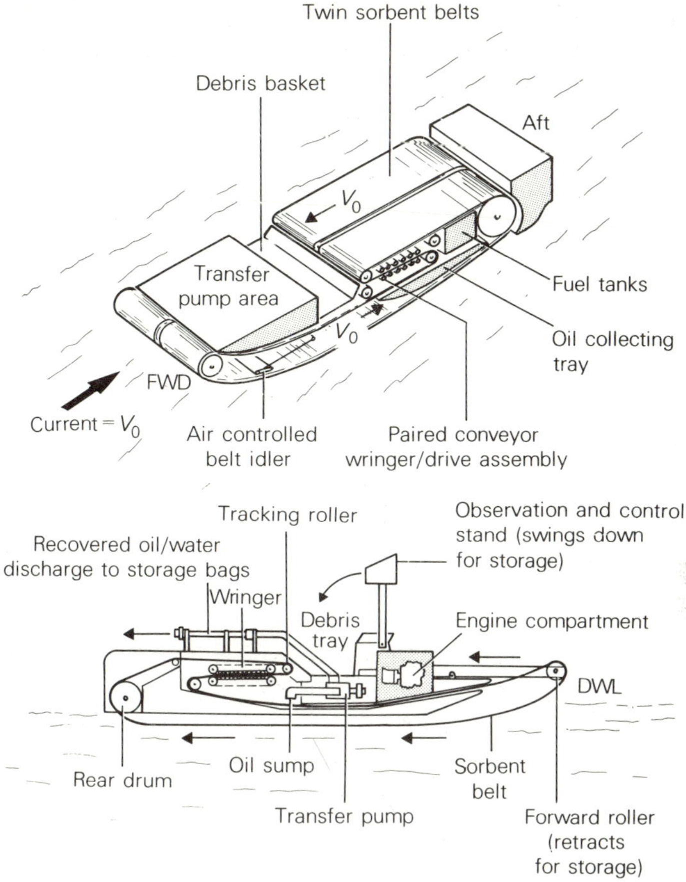

Fig. 6.9 Diagram of Zero-relative-velocity oil skimmer

cations of centrifugal separators to separate oil from water. In general these suffer from low throughput and high power consumption. Nevertheless, centrifuge devices have been built that operate successfully on skimming ponds and other quiescent waters. However under moving water conditions the vortices that are created, tend to be somewhat unstable and result in the collection of little oil with much water, and usually also to the promotion of water-in-oil emulsions within the pumping system.

One of the interesting variants of centrifugal devices is the Bertin skimmer. Basically this consists of a large propeller or paddle wheel, on a vertical axis which, with the aid of floats is arranged to lie below the water surface. Rotation of the paddle forms a vortex in the water and the oil layer thickens in the centre. At this point a suction take-off removes the oil. In the absence of waves or currents it works well, and oil with a very low percentage of water is obtained. It is suitable only for highly protected situations, such as oil recovery ponds.

The Cyclonet[6] consists of a pair of cyclones, one either side of the recovery vessel. It uses the forward motion of the vessel to induce vortices in the surface water as it passes through a slot in the side of the device. Water is discharged from the base of the cyclone, and oil accumulates at the top. From time to time the oil can be pumped away (Fig. 6.10). By reason of its fixed attachment to the sides of the vessel, the Cyclonet does not conform to the sea surface. Thus waves can break more readily over the top of it and carry oil away from it. In addition, the swept width is very small. Deflectors in the form of small booms could be used to concentrate the oil, but as these would also need to be permanently attached to the device (and hence to the vessel), they would, similarly, not conform to the sea surface and would therefore be of reduced efficiency. Unless some means can be found to ensure better wave-following characteristics, such devices, which are intrinsically of high inertial mass (including where appropriate the support for it—the ship in the case of the Cyclonet) it is unlikely that they would succeed in recovering much oil in other than very calm waters.

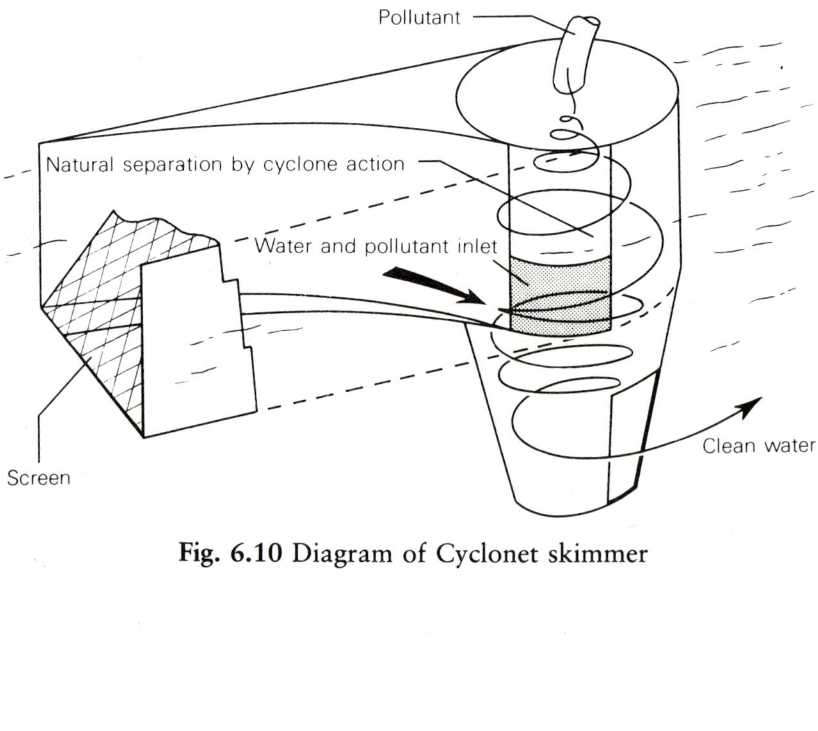

Fig. 6.10 Diagram of Cyclonet skimmer

Group IV: Weir devices

Over a hundred proposals have been made for oil skimming devices based upon the use of weirs. Of these, relatively few prototypes have been built, and fewer still exploited. However, because of their inherent simplicity, low cost and ease of deployment, more weir-type devices are in use than any other type of skimmer.

The rate at which oil is removed by a weir-type skimmer depends on the thickness of the oil layer, the width of the opening, and the speed at which the opening is moved forward.

It is convenient to group and discuss separately those devices that consist solely of the weir mechanism, possibly in a counterbalanced form, from those weir-type devices that are essentially part of a fixed larger structure, such as an oil recovery vessel. The former (small weirs) are suitable only for calm, quiescent and enclosed waters, the latter (larger weirs) for harbour and estuarine conditions. Neither can be considered as suitable for open sea conditions.

Small weirs

Considerable ingenuity has been shown by inventors in manufacturing skimming devices of this general type. Basically, they consist of a horizontal (linear or circular) lip which is aligned, sometimes automatically, with the oil/water interface. Reasonably accurate alignment of the lip with the oil/water boundary is required to avoid collecting a large quantity of water, which would involve an increase in pump size and give rise to a 'downstream' problem in oil/water separation. Many types of counter-balancing arrangements have been suggested to compensate for the effect of waves, but none of these is really successful.

The simplest devices of this kind, such as the Megator, consist solely of a circular weir mounted on floats which maintain the edge of the weir at the oil/water interface. The suction pipe (Fig. 6.11) runs downwards beneath the water and up again to a pump. Simple devices of this type fail if the oil is very viscous as it accumulates on the edge of the weir and does not flow freely into it. However providing the oils are not too viscous, inverted open funnels connected to suction pumps can sometimes be effective.

Increasing sophistication in the adjusting and balancing of the weir at the oil/water interface, and in the provision of oil/water separation can be used to ensure a more efficient collection of the oil from the surface and the delivery from the pump of oil with the minimum quantity of water. The Rheinwerft (Fig. 6.12), skimmer is a good example. The Cascade skimmer has a circular weir

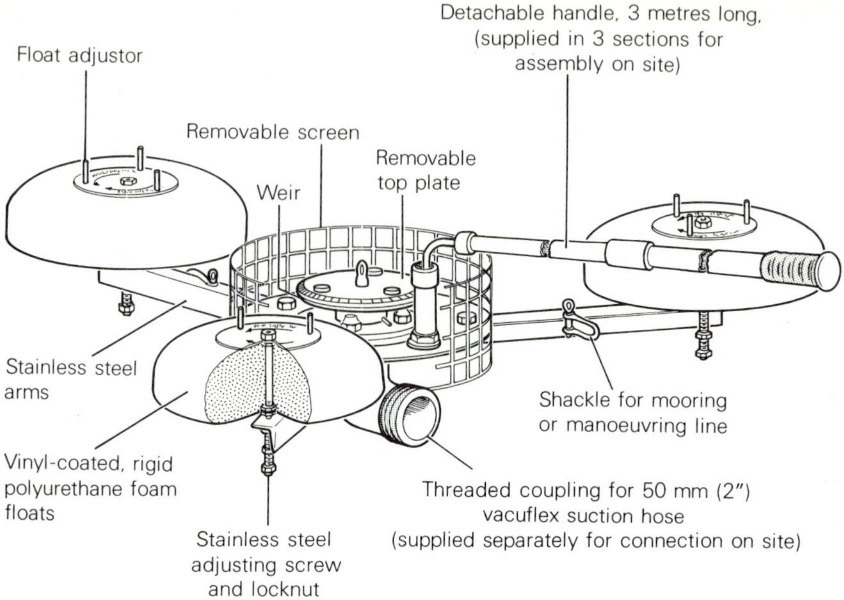

Fig. 6.11 Megator

in the form of a galvanised shallow dish that can be raised or lowered with hydraulic rams. In practice, however, the fine adjustment that can be made to the height of the weir is seldom needed. Like all other devices of its kind, it is suitable for use only on skimming ponds or on calm, sheltered water.

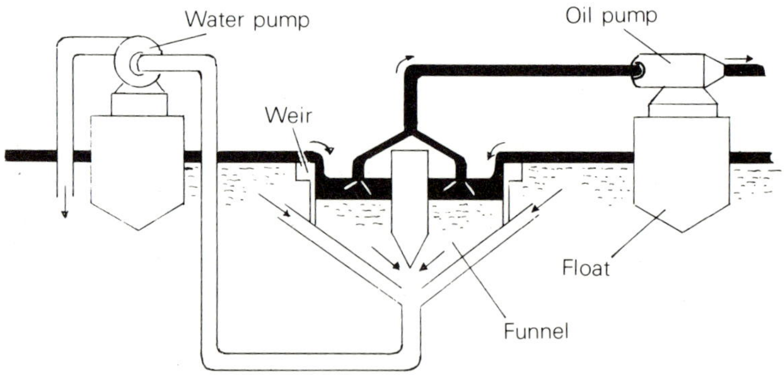

Fig. 6.12 Rheinwerft

An ingenious, but simple, self-balancing weir skimmer of the linear type is the Slurp (Fig. 6.13),[7] developed by the Esso Research Centre. It is easily handled and has no moving parts. It can be operated to recover oil with the minimum quantity of water by adjusting the rate of pumping; this alters the angle of flotation and hence the cut-off level of the weir. The Slurp can operate in water down to a minimum depth of 10 inches, but in approaching this depth, however, care must be taken to see that the trailing hose does not interfere with the balance of the unit in the water.

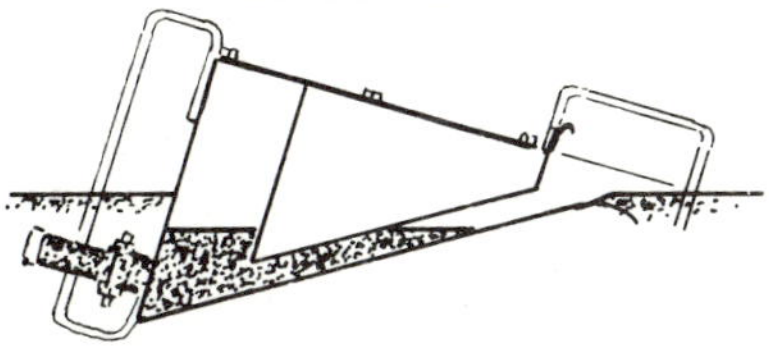

Fast pumping gives a deep weir immersion suitable for the rapid recovery of thick oil slicks.

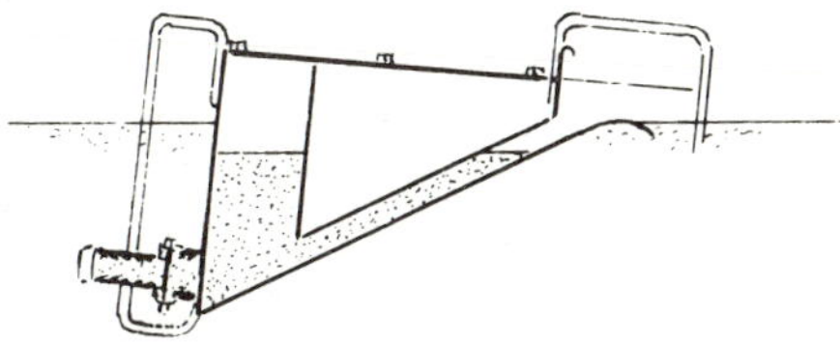

As the slick thins out the pumping rate slows down. The liquid level inside the slurp rises and the skimmer tilts to reduce the weir immersion.

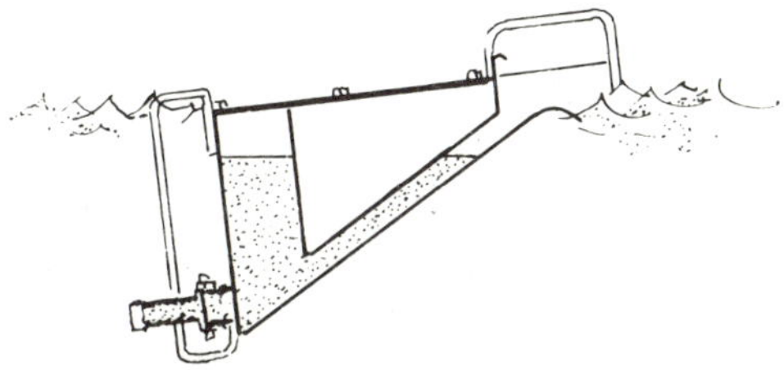

In choppy waves the weir automatically rises until only the wave crests are collected at the rate set by the pump.

Fig. 6.13 Slurp

An overall maximum recovery rate for a free-flowing, low viscosity oil was determined at 68 litres (15 gallons) per minute; above this rate considerable quantities of water were pulled in over the weir. In practice, an oil recovery rate of about 30 litres per minute is considered to be reasonable. To achieve this an oil thickness of about 15 mm is required.

The oil recovery rate falls off markedly at low oil-film thickness, and it may be preferable to set the pumping at a fixed rate to remove both oil and water, and to pass this effluent through a simple oil/water separator to remove the oil before returning the water to the river, harbour, etc.

Tests with paraffin (kerosene), light diesel oil, gas oil, domestic heating oil and crude oil (Kuwait) showed that the Slurp could be used to clear completely the surface of the test tank of oil. With a heavy lubricating oil, and naturally weathered crude oil, the oils were too viscous to flow over the weir, and tended to build up at this point. This gave rise to excessive water being pulled into the pump over the weir under the oil layer, and also to the effect known as 'porpoising' of the Slurp, which again caused excessive water to be pumped away.

In practice the Slurp will skim oil from a rippled surface fairly well, although at a reduced rate. As the wave size increases, the weir is lifted above the surface, porpoising is induced and large quantities of water are then collected.

Debris tends to build up on the trash grid, but unless it blocks the ingress over the weir (this has happened with plastic sheeting) does not seriously affect the efficiency of the device.

As with all small weir type of skimmers, the Slurp is not suitable for operation under open-water conditions of wind and wave. It is, however, effective at clearing free-flowing oil from inland and sheltered waters. Being light, portable and inexpensive, it can appropriately form part of an emergency kit for the cleaning of small spillages of oil on rivers and in docks. Extensive use is made of the Slurp in clearing oil from skimming ponds.

Large weirs

This group contains skimmers which are basically in the form of a catamaran with some type of weir system fixed between the two hulls. Such equipment was a logical development, particularly suitable for the comparatively calm waters of the Gulf of Mexico and the major American rivers. Some of the vessels that have been built are 30 m long with 10 m between the twin hulls. They are not suited to the task of oil recovery at sea, and the stability of some of them as vessels under open seas conditions must be suspect.

Such vessels are usually self-powered and self-propelled. The sweep width tends to be narrow, and oil does not always feed satisfactorily between the hulls. Attempts have been made to increase the sweep width by the use of extension arms in the form of booms extending forwards and outwards from the bows of the two hulls to give a 'V' formation. However with these in place, small waves tend to be reflected from the booms and to build up a choppy sea pattern in the apex of the 'V', close to the weir. Under these conditions, oil will not be presented to the weir in the form of a coherent surface layer, but will be distributed within the top layers of the water.

It was quickly found that a single weir allowed a great deal of water to pass. Consequently, additional multi-weir systems have been developed. The basic idea is that a first weir collects oil and water which then flows into a central holding area where the waves are much reduced and the oil layer tends to stabilise. A second weir permits the oil to accumulate free from water, while the separated water is either pumped out or, more frequently, passes out through holes in the bottom of the vessel.

One of the more interesting skimmers of this type is that known as the Harbor Sweep Skimmer,[8] developed by Shell Pipelines Laboratory at Houston. It is a modified weir type in which the oil is collected by allowing the water underneath to run through holes in a specially shaped and channelled plate. This is mounted between the hulls of a catamaran, the forward end of the opening being filled with horizontal tubes and an 'egg box' device intended to suppress waves. In practice, this catches debris which may have to be removed by hand. The machine has been used for various spills in the Gulf of Mexico and the Houston Ship Canal. An oil recovery rate of 95% at a forward travelling speed of 0.84 m/s (2.75 ft/s) is claimed. In its original form a 'V'-shaped 6.1 m (20 ft) wide diverter was employed ahead of the vessel to increase the sweep width, with a system of flexible, articulated baffles to reduce the turbulence in the collecting area. It is understood that this has now been removed to give a better collection efficiency, although at the expense of the sweep width.

No equipment of this type has been made in the United Kingdom, and very few objective tests have been recorded. Many of the claims made by the manufacturers in their brochures are unsubstantiated and it is very difficult to judge if the devices really work under even mildly adverse conditions.

Group V: Other devices

Ocean systems oil recovery barrier[9]

This is an ingenious combination of boom array and both primary and secondary weirs (see Fig. 5.11). The individual units, about 1.25 m in length are joined together to form booms of up to 30 ft in length that are towed slowly towards the oil slick. The oil, together with some water, passes over the primary weir into a compartment made of an oil-resistant fabric which provides a quiescent area in which waves and water currents are much reduced. A secondary weir then allows the oil to accumulate in a second compartment from which it may be pumped away.

Considerable attention has been paid to the design and shape of the various sections and of the flexible bottom of the first compartment area to ensure that, as the whole system is gently towed through the water and the oil slick, an area of negative pressure is produced beneath the bottom of the first compartment, so that the water below the oil in this compartment flows gently through holes in the flexible membrane.

At the time of compiling the first edition of this book, this system was at the prototype stage. Subsequent tests on a 125 gallons per minute model have not been encouraging, with poor recoveries and loss of oil. Further developments are awaited.

JBF 'SHOC' and 'DIP' skimmers[10]

These are small vessels with an inclined plane extending through the air/oil/water interface, between vertical plates that confine the oil. The motion of the vessel in a forward direction forces the oil downwards along the inclined plane towards a collection well in the afterpart of the vessel. In the DIP skimmer, the oil is assisted in its downward passage towards the collection well by the movement of the inclined plane, accomplished by the operation of an endless belt over rollers. Thus the DIP skimmer does not rely on the forward motion of the vessel for the collection of the oil.

As with some of the devices based upon weirs, the entrance to the collection well is in the form of a grid construction, known as an 'egg box'. This serves to reduce water turbulence in the collection area and helps to form a coherent oil layer, from which the oil can be pumped using a simple weir inlet device. Debris can be collected on a simple trash grid but under certain circumstances it can also build up in front of the skimmer. Oil-soaked straw and seaweed can be allowed to pass into the collection well, although some will get caught on the egg box structure. Difficulties have arisen in transferring weathered oil mixed with seaweed, straw and other debris from the collection area, using the pumps originally provided for oil transfer, but this is not regarded as a serious limitation, and these devices are among the more successful in handling this type of oil-soaked material.

Both SHOC and DIP skimmers are designed for harbour use, and the vessels constructed are clearly unsuited for off-shore operation, although they have been used under choppy conditions where the pitch and heave allowed.

An interesting development is to employ one of these skimmers self propelled, but remotely controlled from the deck of a large vessel such as a tanker. The control cable umbilical also includes the oil take-off pipe. It is claimed that this device could be used in sea conditions much too severe for other weir types.

In Chapter 5 the use of a boom to collect or 'thicken' floating oil has been discussed. This renders the oil easier to collect and remove from the water. A special skimmer, the 'Weir Boom' was described (p. 104). Providing the conditions in the trapped oil are suitable then many of the devices discussed in this chapter could also be used. The Vikoma was specifically designed to use within the BP Vikoma boom. While a skimmer of the Marco type can be used to pick up the oil, either directly within the pool of oil or between two boom sections.

A number of publications give long lists of skimmers culled from the pages of scientific journals, technical and trade literature. Such lists, useful as they may be, give little guidance on which skimmer would give the best performance in a given oil spill situation. To retain their value such lists must be continually updated in order to delete those devices which have not reached, or are unlikely to reach production, as well as to record changes made to production models following actual use. Even knowing what is available does not make it possible to recommend a particular skimmer-even for inshore and calm water use.

The Warren Spring Laboratory has been testing oil spill clean-up devices for over 15 years. The experimental programme of the last few years has involved testing different skimmers on the high seas. This work has enabled the laboratory to draw up a series of guide lines detailing the characteristics a skimmer must have if it is to perform satisfactorily in various sea states. This work has led to the development of the Spring Sweep System.

This experimental programme has clearly indicated that the performance of such devices in general falls off extremely rapidly with worsening seastate.[11] In the case of some devices the oil recovery is negligibly small even in conditions that can only be described as calm. The most severe restrictions on performance are found to be associated with those devices which are intended to be fixed to the sides of recovery vessels. Next come smaller dedicated vessels which incorporate the skimmers unit. Best of all are isolated floating skimmers.

It is quite clear that these results have their explanation in terms of inertial mass and linear dimensions of the skimmers in relation to wave length.

A floating object at sea can do either of two things. It can remain at rest in space while the waves pass by and in so doing the freeboard varies with time, or it can remain at all times in the air-sea interface, rising and falling precisely in phase with the waves. In practice, of course, floating objects behave somewhere between these two extremes and move up and down on the waves but not quite in phase. The extent of the out-of-phase movement depends on dimensions and inertial mass, particularly the latter. On the other hand, oil floating on the sea surface is entirely in phase with the waves. Skimmers will only collect oil at sea in a satisfactory manner if they remain in the oil at all times. To do this, they must be of low inertial mass and of small dimensions compared to wave length.

It is quite clear that if the collection element of the skimmer is above the oil, i.e. in the air part of the time, and below it, i.e. in the sea, for the rest of the time and only remaining momentarily within the oil layer in the manner described, recovery rates will be low. There is however another very serious phenomenon which comes into play as soon as the skimmer begins to move relative to the interface. This additional phenomenon is a variation in displaced water volume with time, i.e. sometimes water is rushing in to fill space beneath the device while at others it is being forced away as the skimmer descends. This action gives rise to turbulence which in turn may prevent the oil from even approaching the device, thus further reducing the likelihood of recovery of the oil.

These aspects of the problem, which have been repeatedly observed, account for the low and rapidly decreasing performance of oil recovery equipment in the presence of waves. Furthermore, even very small waves can have these effects if the inertial mass of the device is high enough.

This being so, special vessels incorporating the skimming function as an integral element in their construction will be inoperative in waves. Furthermore any attempt made to reduce their size in an effort to overcome the inertial mass/linear dimension problem will result in a vessel too small to have any realistic storage capacity for recovered oil. Specially-equipped vessels of this type, furthermore, cannot be deployed quickly. It is preferable, other things being equal, to transport equipment by road or air which can be used on suitable ships already close to the scene of the incident which will preferably be tank ships.

The Springsweep System

The Springsweep System is best described as an example of a system designed for oil recovery, taking into account those features which limit the use of booms as collecting devices as discussed in detail in Chapter 5, together with the extensive experience gained at Warren Spring Laboratory in the course of the evaluation of commercially available skimming devices mentioned above.

The system consists of four components: a boom, a skimmer, a pumping system and a means of deployment of the other components. The whole system can be mounted on a small coastal tanker which provides the receiving facilities for the recovered oil. (See Figs 6.14 and 6.15).

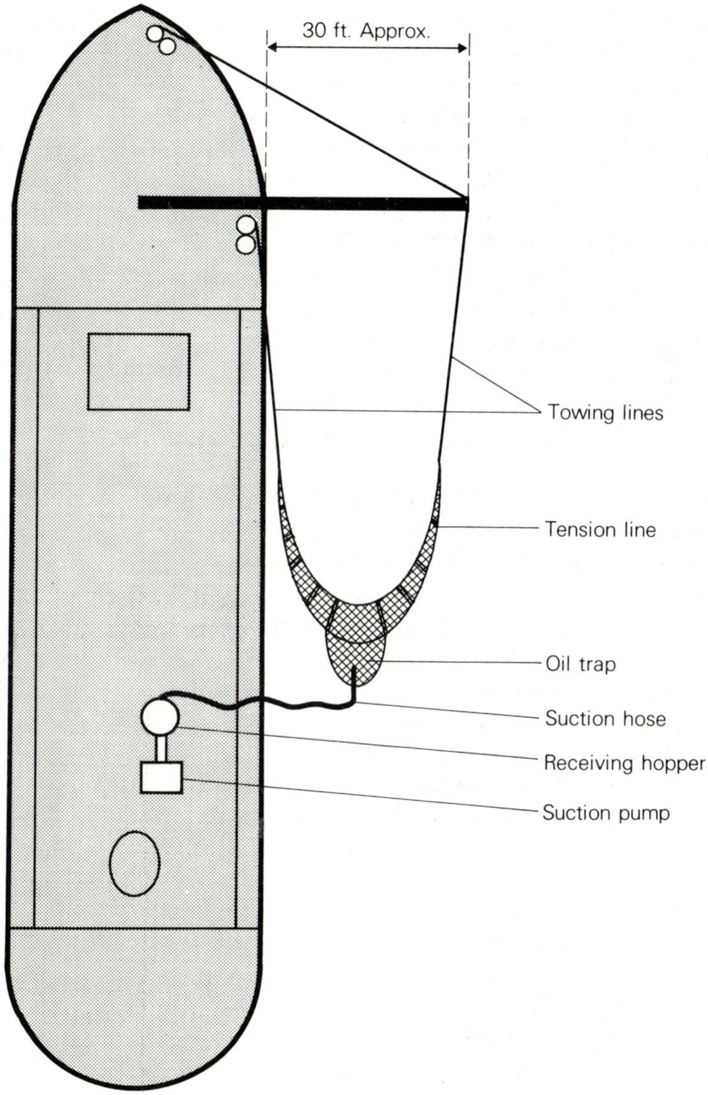

Fig. 6.14 Diagram of Spring Sweep system

Fig. 6.15 Photograph of Spring Sweep system

The Boom: Any separate tension boom which is constructed according to the ideas expressed in the section above, can be used. In the experimental work leading to the definition of the system, the boom employed was a Trelleborg Troilboom, (see p. 113) and this is now supplied as part of the Springsweep System.
The Skimmer: The overriding principle is that the skimmer must be of low inertial mass. It has been found, in practice, that elaborate and sophisticated devices, such as some that are described earlier in this chapter, have little or no advantage over a simple tube whose open mouth can be aligned in the oil layer. For the essential truth of an oil recovery operation on the high seas is that there is little point in the exercise unless oil is collected within the boom pocket at such a rate as to enable it to be pumped directly from the sea through a reasonably sized (say 10 cm) hose.
The Pumping System: The oil flow line must be capable of transferring oils of a wide viscosity range as well as water-in-oil emulsions and debris of all kinds including rags, occasional dead birds etc. Trials of a number of commercial pumping arrangements have indicated that only a vacuum system performs adequately with this wide range of oils and emulsions. Debris can be recovered from a catchment pot.
Deployment of the System: The Springsweep system provides also

the means of deployment and recovery of the components of the system. It should be noted that the sweep width of the system is a great deal smaller than that usually proposed for high seas operation. This is in line with the earlier observations that there is little point in sweeping sea areas where the oil layer has thinned to less than 1.0 mm. The Springsweep system is intended for deployment in those areas where reasonable quantities of oil exist, particularly in windrows. For this situation, the sweep width of some 30 ft. is considered adequate. In summary, the formation of windrows is seen as a natural mechanism for the concentrating of oil that is spilt at sea, and the Springsweep system is one that takes advantage of this. It can of course be used in any situation at sea where the oil layer is thick enough to make a worthwhile oil recovery.

It is interesting to find that the Shell Development Company,[12] Houston, Texas, have, from very similar reasoning, developed a skimmer which they have tested in seas up to 8 ft wave height. This is called the Sock skimmer. It consists of a covered boom which is towed alongside a tending vessel. The boom collects oil in the usual way, and has the same limiting speed, before oil is lost beneath the boom. The cover damps out small waves within the boom, and also eliminates wave reflections from the sides. Elimination of small waves and their reflections is said to remove a major source of oil loss from the boomed area. The draft of the boom is 4½ ft so that a reasonable volume of oil can be collected. The whole system is made of oil resistant fabric which enables the whole unit to be folded up for transport and to readily unfold in the water. Oil is recovered from multiple suction points located in the top cover. The kit includes a power pack consisting of a diesel engine driving a hydraulic power unit and a positive displacement pump. The hydraulic power is used to operate the launching and recovery apparatus and to raise and lower launching ramps which are carried by the vessel. Figure 6.16 shows the equipment in use.

Ancillary Equipment

None of the skimmers described above has capacity large enough to contain all the oil that is, or should be, recovered from even a medium sized spill. There is, thus, the need for ancillary equipment to transfer this oil to a storage vessel prior to recovery ashore or other ultimate disposal that may be on a site remote from the spill or the oil recovery operation.

Fig. 6.16 SOCK skimmer in use on the starboard side of a 190 fast oilfield work/supply vessel in the Gulf of Mexico in 1979

If the recovered oil has to be piped any considerable distance or raised in level, then additional transfer pumps will be needed and arrangements made to connect these various pieces of equipment together. As most skimmers remove water as well as oil, separators may be needed, not only to recover the main body of oil from the water (a settling tank can be used for this), but also to clean the residues of oil from the water before it is returned to the sea. Temporary containers may be required to store the oil before further transhipment. These various devices and the problems of deployment and logistics are discussed in the following sections.

Transfer pumps

For inland spills or oil recovered from inland waters, the use of vacuum trucks ('gully emptiers' or 'sludge gulpers') can be used to pump, hold and transport the recovered oil. Although this principle can be used at sea, the technique is more difficult to operate and it can only be recommended for very small vessels.

In almost all cases where the need arises to transfer oil at sea, an oil transfer pump will be used. The required characteristics of the pump will vary according to the situation, but the principle factors that need to be considered are:

availability and transportability;
power requirements;
priming (should be self priming);
adjustability of throughput;
entrained water emulsifying characteristics;
sensitivity to debris;
effect of oil on the pump valves, linings, etc.

In the main these are characteristics that are pre-determined by the design of the pumps that are available. The same considerations and limitations apply also to the remaining parts of the oil transfer system, including hoses, couplings, separators and storage containers.

One common fault is to use pumps that encourage the dispersion of oil by the formation of fine droplets. This increases the difficulty of making a subsequent separation, increases settling times, introduces the need to handle much larger quantities of oil/water mixture and increases the risk of oil failing to be collected and therefore subsequently being discharged in the ballast water.

The principal factors contributing to the formation of fine droplets during pumping are excessive pumping speed, and excessive fluid velocity. One of the most suitable pumps is the hypocycloid or 'Mono' pump. Double-diaphragm pumps are also suitable for use with small skimmers and experience with the Spate range of pumps has been found to be encouraging. In an emergency, almost any pump, however far from ideal, can be pressed into useful service provided it has sufficient capacity.

Hoses, couplings, valves

Little difficulty is usually experienced with hoses for the smaller skimmers, except that if recourse has to be made to existing hoses, then those available are sometimes too heavy for the skimmer and can upset its balance and hence destroy its performance.

It is almost impossible to press available hoses into service for use with the high capacity oil skimmers: these devices must be equipped with their own permanent allocation of hoses. Ideally, they should be tough, lightweight, oil resistant and easily handled. Lay flat tubing of soft PVC with reinforced polyester yarn has been found to be closest to the ideal, but it can only be used on the delivery side of pumps.

Couplings are a source of impedance to the oil flow, and should be used only where necessary. Debris can also be held within a piping system by the slight restriction of a coupling. Similarly, the imposition of valves or stopcocks in an oil line can give trouble from entrained debris. Nevertheless both valves and couplings are frequently required in oil recovery operations, and care and thought must be given to provide enough of the right sizes and design.

Water separators

The manufacturers of a number of skimmers claim that their devices collect only minimum amounts of water. This claim is not always valid for the circumstances under which it may be necessary to deploy them and, in any case, the use of the device at its maximum oil recovery rate may produce a mixture that contains a great deal more water than oil. In order to make the most use of whatever recovered oil storage capacity is available for such skimmers, some form of oil/water separator is desirable.

In its simplest form, a tank with a stopcock and drain to remove the lower water layer will suffice as a separator. Tanks with bottoms open to the sea are not satisfactory, as oil globules that do not settle will be swept away.

For use with the smaller weir-type skimmers in recovering relative small quantities of oil (gallons rather than barrels), an effective separator can be constructed from an open oil drum, perforated at the base and filled with an oil absorbent such as polyurethane foam. The oil is absorbed on to the polyurethane foam and water is discharged through the holes in the base. Such devices do not work if the oil contains even small quantities of dispersant.

A number of floating oil/water separators have been described for use at sea. A simple device of this kind by Bennett Pollution Controls consists of an open-top conical fabric bag with a float ring at the top and a foam filter at the base. Oil/water mixture is fed in at the top, excess water passes through the foam filter and the oil layer is removed from time to time to storage.

With all separators, the separation of oil is encouraged by maintaining a quiescent pool and by avoiding turbulence or mixing of any kind. If fine droplets of oil have been produced, as for example if a centrifugal pump has been used, they may never coalesce and separate into an oil layer under conditions existing at sea.

Temporary oil containment

In the consideration of any oil spillage contingency plan embodying oil recovery, some thought needs to be given both to the ultimate disposal of the oil, and to its storage for as long as is needed to complete these arrangements.

Oil recovered from rivers and harbours can usually be stored in tanks or tank wagons. Portable plastic swimming pools also have been used and, where such tanks are not available, lengths of polyethylene tubing sealed or tied at the ends. A temporary storage facility can often be created by excavating a depression in the ground and lining it with polyethylene sheeting.

Such makeshift containers are not suitable for use at sea, and if dumb barges or small coastal tankers are not available, then recourse must be made to floating flexible containers such as Dunlop Dracones or the various oil bags produced specifically to contain recovered oil. Such dracones and oil bags are expensive, need a skilled team to deploy and use them, and are available in only relatively small sizes (Dracone to hold up to 360 m^3, oil bags to hold up to 700 m^3).

So far, only limited experience appears to be available in the use of these temporary oil storage facilities, and little is known about their operation in conjunction with high capacity skimmers, their towing capabilities under filled or partly filled condition, and the problems of transferring the contents whilst still at sea.

Deployment and logistics

The recovery of oil from the surface of water, and especially on the high seas is a skilled operation and to be effective requires the use of a trained and well-rehearsed team. It is important for them to know not only what equipment is available, but also what are the factors that govern the choice between available items, how to deploy them effectively and how to continue to make an oil recovery as the conditions change.

Thus, in any incident, it is important to know or to be able to ascertain the type and characteristics of oil involved in the spillage, the quantities involved, the position, drift and spreading of the slick as well as such factors as currents, tides, depth of water in the area, location of shoals and obstacles to shipping, etc.

It must not be forgotten that much of the equipment used for oil recovery, ranging from high capacity skimmers to pumps and Dracones, is still not far from the prototype stage and is not readily portable. If it is not immediately available on board a suitable vessel, then dockside crane facilities may be needed to lift it aboard. Some kind of shipboard crane or derrick will be needed to deploy the equipment at sea.

Oil recovery tends to be a protracted operation. Provision must therefore be made for extensive back-up facilities both at sea and ashore to provide relief arrangements for both men and craft. The importance of good communications cannot be over emphasised.

Although a great deal of time and thought (and money) have been spent on the design and construction of mechanical equipment for the recovery of oil from the surface of the water, experience in its use is surprisingly little and unknown. Most of it has been obtained using prototype designs or equipment under development. There is a need for detailed 'in depth' studies of the use of such mechanical devices, including that of the necessary ancillary equipment and of the logistic considerations.

References

1. BRUCK, B. and MAXWELL, K. R., 'Lockheed Oil Spill Recovery Device', Proc. Joint Conf. on Prevention and Control of Oil Spills, API, Washington, 1971, p. 339.
2. BEYNON, L. R., 'Clean up of oil spills at sea and on the coast', World Petroleum Congress, Tokyo, 1975, PD25 Paper No. 4.
3. BROCHURE, Marine Construction & Design Co. (MAMCO), Seattle, Washington. USA.
4. Oil Mop (UK) Ltd. Tuckton Bridge, Christchurch, Dorset BH23 JS, UK.
5. 'Zero-relative-velocity oil skimmer', *Petroleum Engineer* **49**(3), 11–14 (1977).
6. PINCHON, J., 'The Cyclonet; a device for picking up oil slicks from the sea surface', Proc. Joint Conf. on Prevention of Oil Pollution, API, San Francisco 1975, p. 387.
7. *(Slurp) Clearance of oil from water surfaces*, Slurp skimmer Report LR186, Warren Spring Laboratory, Stevenage, Herts, UK.
8. AYRES, R. R., FRASER, J. P. and KARZIEROZAK, L. J., 'Developing an open sea skimmer', Proc.Joint.Conf. on Prevention of Oil Pollution, API, San Francisco, 1975, p. 401.
9. MARCH, F. A. and BEACH, R. L., 'High seas recovery system', Proc. Joint Conf. on Prevention of Oil Pollution, API, Washington, 1973, p. 339.
10. BIANCHI, R. A., FARRELL, J. H., and JOHANSON. E. E., 'Fixed and moving inclined plane skimmers', Proc. Joint Conf. on Prevention of Oil Pollution, API, Washington, 1973.
11. CORMACK, D., LR 318(OP), Warren Spring Laboratory, Stevenage, Herts, UK, 1979.
12. FRASER, JOHN, P., 'A new oil skimmer', *Ecolibrium* **9** (4), 13 (1980).

7

Absorbents, burning and gelling

Usually, when a liquid is spilt it is mopped up with a dry absorbent cloth which is then wrung out and is ready for use again. It is therefore reasonable to try to use a similar technique to remove oil floating on the water. What is required is something which will absorb the oil and not the water, and which can be spread over the floating oil, removed, wrung out, and used again. Under suitable conditions, and with correct materials, this method is undoubtedly one of the most effective for completely removing spilled oil from the water environment. Unfortunately, this does not apply when one is considering large spills on the open sea. This method is only suitable for small spills on enclosed waters, i.e. reservoirs, rivers and harbours.

There are a number of ways in which a material which does not absorb water but which absorbs oil could be used. If the material floats, and almost all of the suitable materials do float, then it could be used as a powder. This might be distributed over the oil slick by the use of some sort of blower, where, hopefully, it would absorb the oil and stop it spreading further over the water surface, the final result being a dense floating mass, rather like porridge. This mass would then have to be removed from the water. Alternatively, it might be considered better to use the material broken up into small lumps, perhaps cubes of a centimetre size. These might be easier to distribute, being less dusty, and they would obviously float rather higher in the water, and might be considered easier to pick up. On the other hand, the oil would take longer to be absorbed by the pores, and stirring, or local agitation, might be necessary. Again, the floating raft of oil-soaked material would have to be removed from the water surface and disposed of. The removal process is easier the larger the pieces of material. With this in mind, a number of firms manufacture 'cushions', that is, rectangular bags, perhaps 50 cm square, and 5 cm thick, which are filled with powder or granulated absorber. The bags are usually made of a porous or open weave material so that both the oil and the water can readily pass through the pores to contact the absorbent material inside. Bags of this kind can be tied together on a rope and trailed between two boats, where they will absorb the oil, and they are particularly useful in cleaning the last traces of sheen on enclosed waters. When the bags have become oil-soaked, they can readily be pulled out of the water, and perhaps passed between rollers to squeeze out most of the oil, and then they can be re-used. A further extension of the same idea, is to use a large drum covered with an absorbent material, or to use an endless belt of oil-absorbent material which passes through the oil film to a roller held in the water; the belt absorbs the oil and then, in the boat carrying the equipment, it can pass between rollers which squeeze out the oil. The belt then returns to the sea. These developments are most closely allied to mechanical methods for removing oil from the water surface and consequently were discussed in Chapter 6.

If absorbent materials are to be used at an oil-spill, then there are a number of factors which must be considered when making a choice of the most suitable material. Many of the materials are only suitable for use on one occasion. For example, a material such as straw cannot be cleaned or the oil squeezed out so it has to be disposed of after the first use. Consequently, the first important factor is cost and availability. In many cases these are interdependent. Thus in a sugar cane growing area, the dried sugar cane stalks (bagasse) are available at almost no cost. But however satisfactory such material is, if it has to be transported by road or sea for a great distance, its cost would be quite high, and a less satisfactory but locally available material, may well be more cost-effective.

Whatever the material, it has to be distributed over the floating oil and agitated so that it absorbs the maximum amount. How this is done will depend both on the sort of material and also on the means available. In a small harbour, situated in an agricultural district where straw is available, there will be small boats and agricultural implements used for handling straw, such as pitchforks and rakes. On the other hand, in an industrial harbour it may well be necessary to design, and have standing by, special equipment.

The oil-soaked material must be landed ashore and dealt with. Some of the materials can be passed through rollers, and a good deal of the oil, usually accompanied by water, squeezed out. In general, if the oily material is piled in large heaps, particularly if the weather is warm, then oil will almost inevitably run out so that great care must be taken as soon as the oil-soaked material has been landed on shore. The problem of disposal of this oil-soaked debris is discussed in Chapter 9.

In the light of the considerations given above, it will be seen that the use of absorbents is not considered as a universal method. Their particular value is to remove the last traces of oil, which is important in the case of spills on inland waters, and is also essential if the water is to be used as a source of domestic or industrial water. The use of absorbents is indicated in the following situations. For example, if there is a small spill in a river, then a boom made of an absorbent material can be put across not only to contain the oil but also, providing the spill is not too great, to prevent the oil from moving downstream and, at the same time completely collecting and removing it from the river. Similar conditions can easily arise in harbour spills or in coves and estuaries around the coast.

Suitable materials

Oil is collected in two main ways using the type of materials described above. First, oil adheres to the surface or is absorbed on to it due to the surface properties of the material. Theoretically, this is a monomolecular layer, but in actual practice the layer is relatively thick and is limited by the viscosity of the oil; the more viscous the oil the thicker the layer that will hold on a given surface. The second method by which the oil is held, is by capillary attraction, that is, the oil fills up holes or apertures in the material and the capillary forces cause the oil to flow upwards into the material. The rate of penetration of the oil into the material is directly proportional to the radius of the capillaries, and is inversely proportional to the viscosity of the oil. The viscosity of oil has a wide range—some two orders of magnitude. This indicates that the efficiency of these absorbing materials, no matter of what type or chemical composition, is likely to vary very considerably with the properties, in particular the viscosity, of the oil. In addition, many of the oils used in normal commerce, particularly crude oils, are made up of mixtures, so that the use of these absorbers may perhaps remove a lighter fraction, while leaving the higher viscosity material still floating. In considering the different materials that can be used, it is convenient to divide them into groups of inorganic, natural organic materials, and synthetic materials.

Inorganic materials

Dealing firstly with inorganic materials, one can consider things like glasswool, vermiculite, or other similar exfoliated micas, organic ash or pumice and the manufactured insulated material, mineral wool or rock wool which can all be used. One of the more successful, but expensive, inorganic manufactured materials is ekoperl. This, like most of the materials in this group, is in the form of fine granules. It absorbs oil of low viscosity very well but with heavy oils, the surface becomes clogged and the absorbent properties are reduced. In normal circumstances, however, the material still floats. All of these fine materials have the advantage that they can easily be applied by tipping them out of bags from small boats, or they can be sprayed over the water surface using mechanical distributors such as are used for agricultural fertilisers. However, their collection is not easy, because they can only be picked up by using fine sieves, or fine nets. Having collected oil, using one of these inorganic materials, it is possible to dispose of it by incinerating the whole mass. The inorganic material will not burn and can be re-used. This, however, can only be carried out in some kind of furnace under carefully controlled conditions; otherwise a great deal of smoke is produced and the material will not be completely clean. In general, it is better to dump the material in a suitable place, where the danger of run-off from the oil is minimal. These inorganic materials are either manufactured or have to be mined and prepared, so that they are really only suitable for use in special circumstances.

Natural organic materials

It is the natural organic materials which have the largest distribution and are widely used around the world. The materials include straw, hay, reeds, sea grass, peat, sawdust and gorse, together with other available local materials such as bagasse or dried palm fronds. These materials all appear to function by virtue of their having oliophylic surfaces and they trap the oil in the mat of criss-cross strands or fibres rather than by actual capillarity. Of all these materials, straw has been the most widely used and is probably the most successful. In many areas, it is already available in bales which can be broken up and forked onto the oily water, or fed through straw mulching machines or the machines used to collect grass and blow it into trailers or tanks preparatory to converting it into silage for cattle food. The natural organic materials then have to be picked from the water surface using rakes or sieves. Many of these materials absorb water slowly so that if they are left lying in the water, they will gradually float lower and may finally sink altogether. It is, therefore, essential to remove the materials from the water as soon as they have been used, rather than to leave them to collect more oil. Even so, the volume of oil which adheres to a scavenger of this type is limited. The amount collected depends on the type of oil, and other factors, and in the case of straw, varies between 8 and 30 times the weight of straw.[1]

In view of the large proportion of thinner oils which can drain out of the materials when they are lifted from the water, natural organic materials are most suitable for weathered crude oils or heavy fuel oils. In this case, it is essential that local agitation is used to get the maximum effect. Straw was used extensively to clean the beaches at Santa Barbara where it was applied by blowing it out from straw mulching machines. However, the oily straw had to be collected almost entirely by hand, and so this method is only recommended where adequate supplies of very cheap labour are available. Sawdust, which is a waste product in large quantities in certain parts of the world, appears at first sight to be a most attractive material because it can easily be distributed. It has the disadvantage that it very quickly absorbs water and sinks unless it is first treated with, for example, a silicone to make it less water-absorbing. However, the silicone or similar treatment is not easy and usually involves considerable pre-drying. In some cases treated bark, usually pine bark, is available as free-flowing material; otherwise, it needs to be enclosed in a nylon or porous netting. Peat, too, has attracted a good deal of attention as an absorbent, as very large quantities of it are available, particularly in the northern hemisphere. However, it has a very high water content when it is dug, and it is therefore absolutely essential that peat is kiln-dried to a water content of somewhat less than 30% before it is used. This converts a natural material into something that has been partially manufactured and so considerably increases the cost.

There are a large number of other organic/inorganic semi-waste products which have been used in small quantities where they are available. Among them may be mentioned waste paper, charcoal, leather shavings and various types of dried agricultural waste.

Synthetic materials

The most satisfactory materials are undoubtedly the polymeric man-made materials. These have many different properties, and by alterations in their compositions and manufacture the density and pore size can be varied over very wide limits. Foams of polyurethane, of polyether, of ureaformaldehyde, fibres made of various materials, nylon, polyethylene, etc. and shavings and granules of a wide range of polyester materials have been used. Some of these materials tend to absorb water, as well as oil, so that the oil retention capacity is reduced by the fact that pores are partially filled with water. Polyethylene fibre rovings, however, are almost water repellent, and this material can have very good oil absorption properties. Many of these materials have been tested by Schatzberg and Nagy.[2]

These synthetic materials have a secondary advantage in that they are produced by mixing certain chemicals together. The chemicals effervesce, produce foam, swell and rapidly set, the heat of formation being adequate for curing the resulting foam. This means that it is possible to transport very large volumes of foam quite simply in the form of two or more drums of the chemical. Various companies have, therefore, prepared simple equipment to mix the chemicals, prepare the foam, and then to granulate it and blow it onto the actual oil spill. Thus a single well-equipped truck is prepared to deal with a large volume of spilt oil.

Although not of general use for oil spills at sea, absorbents have a part to play in the armoury of devices to deal with oil on water.[3] Fuller has carried out a careful survey of the materials and has suggested merit ratings for their choice (see Table 7.1).

Burning

As is well known, some hydrocarbon products, for example, crude oil and kerosene, gasoline, petrol, etc. are highly flammable, so that when oil is spilt on the surface of the sea and floats about in a large slick, the suggestion is frequently made that it should be disposed of by burning. This unfortunately, has many practical disadvantages.

As has been said earlier in Chapter 2, some fraction of crude oil rapidly evaporates. This is the lower boiling point fraction, and so it is also the one which is easiest to ignite. Therefore, crude oil, unless it is lighted very quickly after discharge, is unlikely to be fired without great difficulty. At the same time, however, in these circumstances it is very likely that the vessel leaking, or discharging the oil is still very close by, so that lighting the oil floating on the surface may well be tantamount to lighting the vessel itself.

TABLE 7.1 Floating absorbents—criteria for choosing

Material	*Availability*	*Treatment needed before use*	*Storage problems*	*Ease of Application*	*Oil absorption*	*Efficacy when wet*	*Leaks out of oil by water*	*Ease of recovery from water*	*Ease of disposal*[d]
Straw	****	***	***	**	****a	*	*	**	***
Untreated sawdust	****	****	*	****	**b	*	*	*	**
Treated sawdust	***	*	**	****	**b	***	***	*	**
Pine bark	***	*	**	****	***b	***	***	*	*
Peat	***	*	**	***	***b	***	**	*	***
Ekoperl	**	****	**	****	****b	****	****	*	*
Vermiculite	*	****	**	****	*	c	—	—	—
Polystyrene pellets	*	***	*	**	*	c	—	—	—
Polyurethane foam	**	*	****	**	****b	****	***	**	***
Polypropylene fibres	**	***	**	***	****a	****	*	****	****

NOTE: more stars, the better
[a]Less effective with thin oils.
[b]Less effective with heavy oils.
[c]In view of low absorption not assessed further
[d]Mainly by burning.

If, however, the slick is isolated and still relatively fresh, then it may be lit. This could be a hazardous undertaking. At the time of the *Torrey Canyon*, experiments were carried out with devices which lit spontaneously when dropped on to the water surface and which then burnt vigorously for some time, and these were reasonably satisfactory. Subsequently, a German firm produced a substance called Kontax. This material was in sealed polythene bags. To use, the polythene bag was punctured and the bag was then thrown into the sea where it lit spontaneously and burnt fairly vigorously for some time. Undoubtedly, a material of this kind, which can be dropped from the air, provides a much safer way of lighting a crude oil slick then doing it from a boat. Ignitors, dropped from helicopters or low flying airplanes would seem essential if oil lying on ice in arctic conditions is to be dealt with (see p. 219).

Usually, as discussed earlier, oil spreads out over the sea and rapidly forms a thin layer. This fact presents another difficulty in burning the oil. The lower surface of the thin layer is cooled by the sea, so that although the lighter fraction burns off fairly rapidly and the heat of combustion helps to volatilise some of the other products, nevertheless a considerable amount of the oil is left unburnt on the sea, and this resembles a heavy fuel oil. It is this heavy fraction which is the most objectionable when found on a beach, so that attempting to burn the oil is not only difficult but also the final result does not help the situation very much.

Experiments on the burning of oil on water were carried out in the United Kingdom in 1969. The work was done at a Government research establishment of the Ministry of Defence (the Royal Armament Research and Development Establishment), and co-ordinated by a Working Group of the Institute of Petroleum.[4] In these experiments, a concrete lined pond, 100 sq ft in area, was filled with water. Two thicknesses of fresh Kuwait oil, half an inch and one-eighth of an inch thick were floated on the water. Both ignited readily on the application of the flame and burnt fiercely for four minutes and one-and-a-quarter minutes respectively, and then went out quite suddenly. When, however, a layer of fuel oil was used, or crude oil was allowed to weather for twenty four hours before ignition, then it was found very much more difficult to ignite the layer of oil. In all cases, a residue was left amounting to 15 to 20% of the oil originally applied. However, in practical circumstances, it is possible that as little as about 30% of oil will be consumed by burning. These experiments emphasised the fact that a weathered crude oil could not be ignited by the application of the flame after the weathering loss had reached approximately 20% by weight. This is equivalent to the loss of material with a boiling point below 180°C. Attempts were also made to ignite 'chocolate mousse' made from Kuwait crude oil floating on the tank. None of the methods tried succeeded in igniting this material, so it must be assumed that it is virtually impossible.

Although burning the oil on the water surface is not a particularly satisfactory operation, nevertheless there are cases, for example, the *Torrey Canyon* in which it would seem that the tanker containing the oil is firmly lodged in such a position that it cannot be dragged off or the oil pumped out of the ship. Consequently, rather than waiting for the ship to break up under the action of wind and waves, and gradually release the oil to contaminate the environment, it may well be better to try to burn the oil while it is actually in the tanker. On various occasions throughout the world, tankers have caught light and have floated on the sea, burning vigorously. In the case of the *Mildred Brøvig*, the ship burnt for some months in the North Sea, but at the end of the time about 40% of the oil was still remaining, while the *British*

Crown floated in the Persian Gulf for two months, and even after that time, there was a great deal of oil left in the vessel. It will be remembered that in an attempt to release and set on fire the oil in the *Torrey Canyon*, the ship was bombed, the oil was released, and some 20,000 tons of oil were burnt.

The wrecking of the 89,000 ton tanker *Jakob Maersk* at Oporto (see Fig. 7.1.) in January 1975 has put a rather different complexion on the whole question of burning oil in tankers. In this particular incident, the ship grounded heavily and a tank or tanks were ruptured and the oil entered the engine room causing a series of violent explosions. Following this, the tanker burnt very severely for some 36 h. At the end of that time the ship was in three parts, and apparently all of the tanks, with the exception of one bow tank, were empty. During this time the wind was blowing on shore which was only about half a mile away. When the fire finally went out there was little further pollution from the ship. In fact, the whole shore pollution might well have been produced by only 1000 tons or so of oil. In these circumstances, in which the bottom was ruptured and a heavy swell apparently pumped the oil out of the tanks in turn so that it burnt fiercely both on the ship and on the surrounding sea, undoubtedly a great deal more oil was consumed, and left less heavy residue, than would have been expected from examination of previous incidents.

The Rocket Propulsion Establishment of the Ministry of Defence, discussed the problem of burning oil in tanks with the Institute of Petroleum Committee and finally set up small scale experimental burnings at their establishment during 1968/69.[5] In these experiments, model tanks of up to one cubic metre capacity were filled with either crude oil or fuel oil. These tests were intended to evaluate the effects on burning rate of size and position of vents in the tank, of wind speed and direction, and the partial cooling of parts of the tanks to simulate shipboard conditions. The experiments which were based on tanks of three sizes, with sides ranging from 0.3 m to 1.0 m, suggested that combustion could be effectively sustained if the tank were vented at the top and in one side. The burning appeared to reach a maximum when the areas of the top and side vents were each equal to about 10% of the plan cross-sectional area of the tank. If the area of vents was increased, either by one larger vent or several smaller ones the burning rate appeared to remain roughly constant at the maximum value. The burning rate increased with tank size. With vents of about 10%, it appeared to increase in a non-linear manner suggesting that a constant maximum value would ultimately be reached. The effect of wind was shown to be quite considerable.

Fig. 7.1 The burning of the *Jakob Maersk*

With the side vent to windward and a wind speed of about 11 m/s the burning rate was about twice that obtained with still air. In the small tanks, the burning rate remained substantially constant until about three quarters of the oil had been consumed, after which it was rapidly reduced. It was also found that the cooling of one side and the base of the tank, which was done to simulate the actual conditions of a tanker at sea, had only a slight effect on the burning rate.

In the light of these experiments it was decided to carry out one test using a much larger tank.[6] This was a tank with a square cross-section of 6.1 m and 7.3 m high. A hole was made in the roof and a second in one wall above the oil level, each about 3 m by 1.25 m, that is about 10% of the area of the plan cross-section. In October 1970, the tank was filled to a depth of 5.57 m with 205 cubic metres (approximately 175 tons of crude oil). Due to local supply problems the oil was supplied from the British oil field at Nottingham, England. This oil is similar to Kuwait crude oil, the most marked difference being a higher wax content in Nottingham crude oil. The oil was ignited by means of a floating electrically-fired ignition device which burnt for nearly half a minute and would operate in the absence of oxygen. Within a half minute the fire had spread across all the surface visible through the side hole. There was a light rain falling for most of the time. The fire, as expected, produced a great deal of smoke. In fact a Sunday newspaper headline said that a pall of smoke lay across Southern England. After some five hours burning there was an eruption of burning oil which was ejected out of both apertures. This was most unexpected, the temperature distribution in the oil giving no warning of its onset. A similar eruption occurred 15 h after ignition. In this case some 2 tons of oil was thrown about 50 m against the wind, most of it falling in front of the tank. Unfortunately, these two eruptions burnt most of the instrumentation wires running into the tank. However, the fire finally went out after some 37 h of burning when 97½% of the original oil had been burnt, although the residue was still smouldering in two places for some time afterwards.

Tests with the smaller models had produced a tentative rule which suggested that for a given degree of venting, the mean burning rate R is related to the rate $R\infty$ of an infinitely large tank by the expression:

$$R = R_{\infty}\left(\frac{S - 0.12}{S} \right)^2$$

where S = length of tank sides in metres. In the small tank experiments using Kuwait crude oil with the 10% apertures, the values of $R\infty$ were estimated to be as shown in Table 7.2

For a tank of size 6.1 m, R has a value of 96% of $R\infty$ and the burning rate is given by $(76 + 17W)$ mm/h where W is the wind speed in metres per second. This formula was used with the wind

TABLE 7.2: Wind speed and burning rate

Wind speed, m/s	0	6	11
Burning rate, mm/h	80	190	265

vector normal to the side aperture to calculate the burning rate and to compare with the actual measured burning rates (see Fig. 7.2).

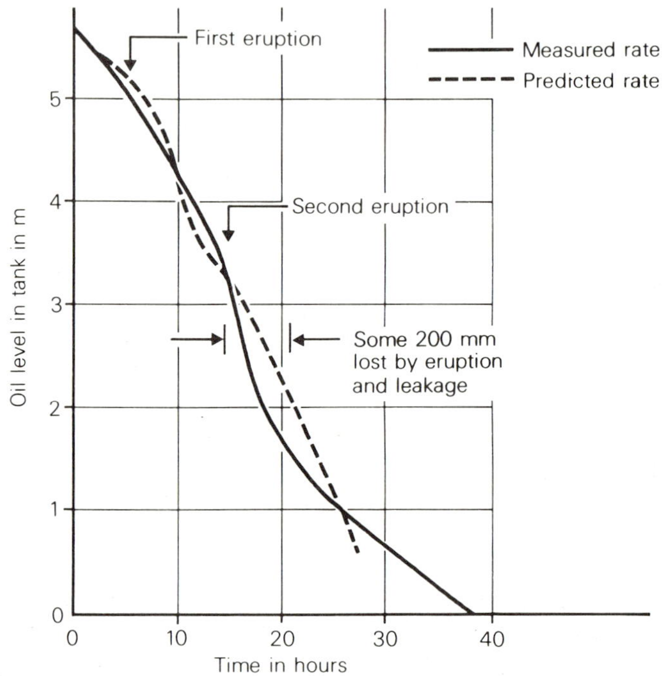

Fig. 7.2 Actual and calculated burning rates

The result of the final test and the model test have enabled burning rates of oil in vented tanks of tankers to be calculated. In view of the difficulties of variations of wind speed due to the natural conditions in the large tank experiment and the difficulties with wind tunnel experiments in the small one, the correlation for extrapolation to cargo tank sizes is given by broad bands rather than by individual lines. However, Fig. 7.3 gives burning rate for three wind speeds against length of tank side. It will be seen that there is a limiting value of the burning rate which is approached as the plan cross-section of the tank is increased. Within the accuracy of measurement this limiting value was reached with a tank with sides of some 6 m.

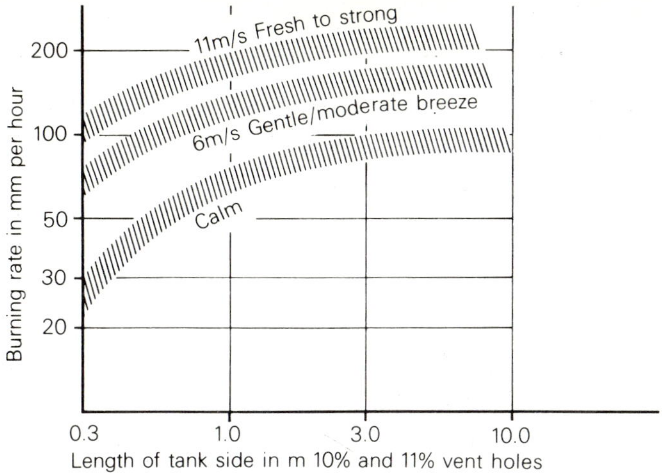

Fig. 7.3 Variation of burning rate with tank size for three windspeeds

The conclusions reached in these experiments are most important. They suggest that, providing certain conditions are met, it is possible to ignite and burn over 97% of the oil contained in a compartment of a stranded tanker. This will be expected to burn at an approximate average rate of 150 mm/h (6 inch/h) and the residue will be of such a consistency it will offer little or no risk of pollution. To obtain these results the tank must be vented at the top and on the windward side, each of the vent areas being approximately 10% of the plan cross-sectional area of the compartment. With the compartments initially full the cutting of the side vents would have to be delayed until a preliminary burn from a hole made in the top of the tank had lowered the oil level sufficiently. Using only top ventilation, the fire would undoubtedly go out of its own accord with a great deal of oil left in the tank. The greater the amount of oil burned the greater is the latitude for creating the side vent holes.

It must be appreciated that in the modern tanker the compartments or individual tanks are very large indeed. They may be 200 ft long, over 80 ft deep, and over 80 ft wide, and can contain up to 20,000 tons of oil. The structural steel work is complex and heavy and only small openings give access to the tank from the deck. It is possible to cut apertures in tanks of this sort by using explosive charges. However, the whole effectiveness of this type of explosive contact charge, depends upon continuous intimate contact of the charge and the steel plate to be fractured or cut. The structural members supporting the deck or side plating could not be severed by this means, so that the openings would have to be confined to deck or side plating panels between the structural members, and the required area would have to be achieved by opening a number of relatively small apertures. The problem of cutting holes in the side after the fire has burnt the oil down to a suitable level, would present grave difficulties and might have to be done by gun fire.

It is now possible to specify how oil could be burned *in situ* in a tanker. But the practical problems are very great indeed, and this method is likely to be employed only in exceptional circumstances. Fresh crude oil floating on the sea surface, can be ignited and will burn, but only if it still contains a high percentage of its volatile content. Even then, burning is unlikely to be complete and substantial residues may remain. Oil which has spread out less than about 2.5 mm thick, cannot be burnt successfully and, in general, it can be said that attempting to burn a floating layer of oil is an unsuccessful method of dealing with oil pollution.

Gelling

One of the reasons for using particular absorbents is that they tend to immobilise the oil and prevent it from spreading rapidly over the water surface. Therefore it would seem to be a possible method to prevent or reduce the spreading of oil if the oil could be persuaded to coagulate or become jelly-like. This method is adopted in the manufacture of non-drip paints, and is also used in military applications when liquid petroleum fuels, under some circumstances, are gelled for ease in handling. If a process of this kind were developed there are two situations in which it could be used. First, the oil floating on the sea might be converted into a gel or thick emulsion to prevent it from spreading, or oil in the tanks of the tanker could be converted into a jelly so that if the tanker wall was perforated the jellified material would not escape.

In general, the formation of a gel, needs the addition of a chemical agent, and a good deal of mixing, which requires a supply of energy. From this fact it is clear that it is impossible to develop any form of chemical which could be used after the oil had actually spilt on to the water surface. It might, however, be possible to gell the oil while it was still in the tank of the vessel. Experimental work, carried out largely by Esso in the USA, produced a compound which, if mixed into oil or crude oil in the tank, would convert it to a jelly so that if the tank was perforated the contents would not flow out. Although this was successful and the quantity of additives was relatively small, the amount of energy required was still quite large and so far it has been found impossible to develop any type of device which could be fitted to a tanker so that, should it get into difficulties, the oil in its tanks could be instantaneously gelled. The alternative suggestion is that all oil carried by sea should be gelled when it is pumped into the tanker, assuming that the action of discharging the vessel can still be carried out without undue difficulty. The additional cost of the additive and of the mixing energy is, however, far too great to compensate for the potential added advantage of preventing or reducing the oil discharged to the sea should the tanker receive any perforation in the tank wall due to collisions or grounding.

Another ingenious alternative[7] which was developed, but which did not prove practical, was to spray the oil with a liquid which was ferromagnetic and miscible with the oil. This enabled the oil to be collected from the water surface using magnectic sweeps. Once more, although this could undoubtedly be carried out in a laboratory test, on a large scale it proved to be impractical.

References

1. UK AEA, *Oil Pollution at Sea*, AERE-R5550, HMSO, London, 1967.
2. P. SCHATZBERG and K. V. NAGY. *Sorbents for Oil Spill Removal*, p. 221.
3. FULLER, 'Use of Floating Absorbents', *Journal of the Institute of Petroleum* **57** (553), 35 (1971).
4. R. MAYBOURNE, 'The work of the Institute of Petroleum Working Group on the Burning of Oil', *Journal of the Institute of Petroleum* **57** (553), 12 (1971).
5. *Burning of Oil in Wrecked Tankers*, Technical Memorandum No. 556, Summary Report on Model Tests, Rocket Propulsion Establishment, 1971.
6. *The Burning of Oil in Wrecked Tankers*, Technical Report No. 72/9, Rocket Propulsion Establishment, December 1972.
7. KAIZER, R., MISKOLEZY, G., CURTIS, B. A. and COLTON, C. K., Recovery of oil from water with magnetic liquids. API, EPA Conference on Prevention and Control of Oil Spills, Washington, DC, March 1973.

8

Sinking and dispersing oil

Sinking Oil

Oil floats as a layer on the surface of water. It would therefore, appear to be possible to get rid of the floating oil by making it heavier, that is, by distributing over it some powdered or granular solid so that the combined density of the solid together with the oil would be sufficient to make the oil sink. A simple experiment carried out with fine sand and oil floating on a bucket of water shows that this is indeed possible. It was at an early stage of the work dealing with oil pollution that proposals were made to deal with it by sinking. The powder has a number of requirements which can be specified. The material itself should be oleophilic so that it will attach itself firmly to the oil. It should also be of high density so that a minimum amount of material is required. The particle size is also important. With a thick layer of viscous oil it is obviously possible to use quite large pieces of the heavy material, even small pebbles for example, but if the oil is of low viscosity and the layer is thin then large particles will drop straight through the oil with very little of it adhering to them. The characteristics of the surface of the solid and of the oil determine how firmly they are bound together. The total surface area of a given mass is inversely proportional to the particle size. Surface treatment can improve the oleophilic properties. Table 8.1 shows the result of some experiments on the amount of powder required to hold down a given amount of oil and, in general terms, it will be seen that the finer the powder the better.[1] The ideal specification can then be summarised as a dense, finely-ground, free-flowing powder.

TABLE 8.1 Crude oil retentivity (24 h) of untreated and treated size fractions of sand

Mesh size BS sieve	*Surface area* m^2/g	Crude oil (ml) retained by 10 g of sand for 24th				
		Untreated	Silicone	Armac T	Duomac C	Duomac T
− 6+ 10	0.09	2.9	3.4	3.1	2.5	0.5
− 10+ 14	0.08	2.9	3.7	4.2	3.4	0.6
− 14+ 25	0.08	3.7	4.7	4.6	3.7	0.8
− 25+ 36	0.10	3.8	4.4	4.8	3.6	1.3
− 36+ 52	0.10	4.7	4.7	4.9	3.8	1.8
− 52+ 72	0.16	5.0	4.3	5.2	4.0	1.8
− 72+100	0.14	5.0	4.3	6.3	4.2	4.2
−100+200	0.29	5.0	4.2	6.3	4.4	4.4
−200+300	0.49	5.0	4.4	6.2	4.9	4.4
−300	0.78	5.2	4.7	6.9	4.7	4.5

One of the earliest commercial attempts to exploit this method of dealing with oil was in the early 1960's when a commercial firm offered an oil sinking service; for a fee they would go out in a plane and distribute a powder over an oil slick and sink it. The planes were small crop-dusting planes with a limited weight lift capacity. The powder was alleged to sink the oil rapidly to the bottom and there to bind it into a solid mass, as the powder, in contact with water, set hard. Although details of the powder specification were never published, it appeared to be a mixture of dense sand, such as barytes, together with a setting material, probably cement. In actual practice, the material was not very satisfactory. Discharging such a powder from an aircraft meant that there was a considerable degree of separation; the finer cement powder was blown away by the wind and the denser powder fell straight down onto the oil slick. It was, therefore, very difficult to direct the powder from the aircraft on to the slick. To obtain the claimed setting, even in a laboratory experiment, a considerable excess of the powder over the amount required to sink the oil had to be employed. As the cost of the whole operation promised to be rather high, and the logistics of dealing with a large oil spill with small aircraft unfavourable, the project was soon dropped.

In France, at the time of the *Torrey Canyon*, another type of powder was developed and was used by the French Government in large quantities. This was called *Craie de Champagne*. It consisted of finely powdered chalk, treated with a small quantity of surface active agent, the same material, in fact as is sold in the building trade in England under the name of 'whiting'. This material, though not very free-flowing, appears to fill the required specification. However, because it is a fine material which is easily carried by the wind, the French found that the only way it could be distributed was by the crew of small fishing boats emptying it out from sacks on to the floating oil. This was clearly a wasteful and uneconomical process. Experiments carried out in England by the UK Government[2] showed that the powder, though it worked satisfactorily under laboratory conditions, was almost impossible to use on the open sea. These experiments, and indeed reports from the French, showed that the powder was likely to accumulate with the oil on the sea bed and could result in the fouling of nets, and of the catch as well, should bottom fishing occur in that zone.

Nevertheless, sinking appeared to be an attractive method. Consequently, the Royal Dutch Shell Company undertook a long programme of research to develop a more satisfactory method of application. In brief, they proposed to use a fine sand treated with an oleophilic agent and then to discharge it, not as a dry powder but mixed with water as a slurry. This proposal had a number of advantages. It was much easier to treat a sand/water slurry with a chemical than to carry out a similar mixing operation with dry powders. The sand/water mixture could be pumped and handled by machinery already developed for the process and its discharge over the oil through suitable nozzles would avoid the problems of the drift of powder in the wind. At an early stage the Company decided that the most satisfactory way of obtaining the sand, and also to distribute it over the oil slick, was to use a suction dredger as the basic equipment. The Netherlands has great experience in dredging operations and there are a number of firms who possess large ocean going dredgers. These dredgers can pick up sand from a water depth of a hundred feet and can then transport several thousand tons of it in specially constructed holds. It was a relatively simple matter to install special mixing tanks so that the chemical could be added to a sand/water slurry and to develop arms, projecting from each side of the dredger, carrying nozzles to discharge the sand/water slurry in a uniform manner from each side. The first work was done using a static dredger. Open barges filled with water, on which was a layer of oil, were towed underneath the curtain of falling water/sand slurry. This enabled accurate measurements to be made of the amount of oil which was sunk with varying types and thicknesses of oil. These experiments proved most successful and the next stage was therefore to fit up full-scale equipment.[3] The dredger used was the trailing suction-hopper dredger *Geopotes VII* owned by Adriaan Volker NV; it is 107 m long with a beam of 17 m and a displacement of 4322 GRT with a loading capacity of 4426 m^3 (Fig. 8.1).

Fig. 8.1 Nozzles on dredger *Gepotes VII*

This work finally resulted in a full-scale trial in the English Channel with the complete co-operation of both the Dutch and British Governments. Some 100 tons of oil was discharged from a small tanker and then the *Geopotes VII* passed over the oil, successfully sinking 95% of it. In the practical case of a massive spill this would be an enormous advantage as the remaining oil could much more easily be dealt with by the use of dispersant. There appeared, however, to be a number of difficulties. It was found that when the site of the sinking was visited on subsequent days, the oil was still rising back to the surface. Therefore, it would appear that in an actual case, as for example, if oil was sunk off an amenity coast with an on-shore wind, then some oil would slowly be returned to the shore by the wind, and the initial success might be considerably reduced. It was alsc found, by repeating the early experiments, that trawling over the area resulted in a fouling of the equipment and of the catch.

It was appreciated from the earlier work that the oil tended to sink as soon as the density of the oil together with the powder exceeded that of the sea, and not necessarily when it had obtained an adequate amount of the powder to retain it properly on the bottom. Observations on the initial experiments made in the United Kingdom by skin divers, showed that the oil descended to the bottom, not as large 'plates' but in relatively small pieces or lumps, and when on the bottom, showed little tendency to coalesce. The skin divers reported that the oil moved in the current like autumn leaves. To examine this point further, experiments were carried out at the Hydraulics Research Station by the Warren Spring Laboratory in 1970. In these experiments a small patch of oil approximately one metre square was sunk in one of the Laboratory's flumes. The sides of the flume were fitted with glass so that the sinking of the oil to the bottom could be observed and also currents of water of varying measured speeds could be applied to see what happened. In brief, it was found that in still water the oil sank to the bottom and was, of course, immobile. However, as the current rate increased there was a low speed, (about 0.2 knots), at which quite suddenly the oily lumps became mobile and were swept away.

As a result of these experiments, the British Government drew up a number of working rules:

1. Oil should not be sunk where there are currents likely to move the sunken oil either in the direction of amenity beaches or of shellfish grounds.
2. Oil should not be sunk where the bottom is likely to be trawled.
3. Oil should not be sunk if it is likely to fall on shellfish beds or fish spawning grounds.

The Dutch experiments indicated that it was essential that the sand was clean and free from clay. The presence of even a small amount of clay increased considerably the amount of chemical that had to be added to produce an adequate coating on the sand.

A great deal of this work was repeated by the United States Corps of Army Engineers who are responsible in the USA for dredging all channels and who consequently possess a number of dredgers.

The general conclusion that can be reached is that, although this is a satisfactory method, it requires the modification of ocean going dredgers and it can only be used where suitable sand is available within a reasonable distance from the site of the incident. *Geopotes VII* stood by in 1970 when the damaged tanker *Pacific Glory* was towed along the Channel, Nevertheless it is believed that this is not a satisfactory method for every situation.[3]

Oil Slick Dispersants

The use of dispersants has been, and still is, the most widely employed and immediately effective method of combating oil pollution both on the open sea, in estuaries and on beaches.

When oil, which is floating on the surface of the sea, is treated with a suitable surface active agent and agitation applied, the oil is broken up into particles of different sizes. Some of the particles formed are so small that they constitute an oil-in-water emulsion, while others are relatively large and may require the application of additional mechanical energy to break them down further. This breaking of the continuous oil slick into small particles is called dispersing the oil. Basically, all dispersants (which are surface active agents in suitable carrying media) work by reducing the oil/water interfacial tension, thus promoting droplet formation. Hence, when a fixed amount of energy W(ergs) is applied to the mixture, considerably more surface area A(cm^2) will be formed when the oil/water interfacial tension $\gamma_{o/w}$ (dynes/cm), is lowered by a dispersant, the relationship being:[4]

$$W = A\gamma_{o/w}$$

Once the oil has been dispersed, the natural motion of the sea is usually sufficient to distribute the particles over a large volume, thus preventing their reforming into a slick. The bacteria in the sea which attack the oil can do so only at the oil/water interface. Hence the smaller the particles, the larger the surface area exposed to the sea and the higher the rate of biodegradation or decomposition of the oil. Thus the use of dispersants quickly removes the oil from the surface of the water, distributing it in small particles throughout the upper few metres in a form which is readily susceptible to attack by the organisms naturally present in the sea.

It is, however, important to recognise the limits of dispersant effectiveness. Different oils exhibit different characteristics: some fuel oils are extremely viscous or even solid at sea temperature whereas lighter fractions such as distillates spread out rapidly over the surface of the water and the problem may well be one of fire hazard rather than one of pollution.

Whilst dispersants normally work satisfactorily on freshly spilled low and medium viscosity oils, they may not be effective on oils of high viscosity. Also, as light, low viscosity oils weather, they increase in viscosity and often form water in oil emulsions or mousses. Thus, they too become difficult to disperse[5,6] recent work indicating that the limit of effectiveness is between 7000 and 10,000 cS at application temperature. In these instances reliance must be placed on mechanical recovery systems.

Dispersants are generally not recommended for use on petrol, distillates and spillages of other light fractions which are often best left to evaporate, due care being taken regarding the fire hazard. However, there are occasions when the use of dispersant even for these light ends is desirable depending upon the location of the spill and this is best discussed on site.

Development of Oil Slick Dispersants

Towards the end of the 1950's there was an awareness that oily patches were appearing on amenity beaches around the coasts and that their frequency was increasing. It was found, at that time, that the industrial version of a commercially available paint brush cleaner, when sprayed onto the oil, effectively softened and dispersed it by means of emulsification as the tide came in and washed over it. It is really from this beginning that the oil slick dispersant stems. At about that time, the United Kingdom Government Warren Spring Laboratory (WSL) started work on beach cleaning chemicals and approved several products for that purpose, all of which were based on highly aromatic solvents and non biodegradable emulsifiers.

Early in the 1960's further experiments showed that it was possible to disperse floating oil before it hit the coast and the products already used to clean beaches were found to be satisfactory. The Fisheries Laboratory of the Ministry of Agriculture Fisheries and Food (MAFF) at Burnham on Crouch[7] at the same time studied the toxicity of these materials to marine life. Initially, the tests used oysters, mussels, cockles and winkles and, later, the brown shrimp. All the dispersant chemicals were found to be highly toxic causing a 50% mortality during a 48 h exposure with concentrations of the order of 10–100 ppm. Nevertheless, these materials were considered suitable for beach cleaning although it was appreciated that some inter-tidal animals and plants would be killed and others, in the immediate vicinity of the spraying, might be tainted. Harm to animals below the low tide mark would, at worst, be very small.

In 1967, as soon as the *Torrey Canyon* was wrecked on the Seven Stones Reef, discussions were held with MAFF to determine whether dispersants should be used. It was agreed that in view of the immense volume of sea off Cornwall and the frequency of the rough seas at that time of the year, the rate of dilution and dispersion would be so great that the risk of serious damage from the use of dispersants could be ignored. The risk to the holiday industry of leaving oil on the beaches was far greater. These views on the harmful effect were substantially correct for, despite the use of 2.5 million gallons of dispersant, no damage was done to commercial fisheries in the short or the long term and the damage to the inter-tidal zone, though severe in places, was completely restored in one or two years.

A great deal more research to produce a low toxicity oil dispersant was subsequently (after 1967) carried out and BP 1100 was the first of this type to appear on the market, consisting essentially of a hydrocarbon solvent with low aromatic content and a biodegradable low toxicity emulsifier of the natural fatty acid polyglycol ester type. Thousands of gallons of this type of low toxicity oil dispersant have been and still are being used most effectively for the dispersion of oil at sea, in harbours and for the cleaning of beaches.

In 1975, the first of the so-called third generation of dispersants appeared on the market, this presenting a new approach to the problem. The product, of low toxicity to marine life, was in concentrated form and had—in order to be used with the conventional ship spraying system—to be diluted at the rate of 1+9 with sea water just prior to application. It thus increased the spraying time of the vessels for a given pay load by a factor of 10. These concentrates are now in general use for off-shore work but the hydrocarbon solvent type products are still often preferred for beach cleaning. Concentrated dispersants, if used undiluted, may require the application of little or no external energy to produce good dispersions. They have thus been called 'self mixing dispersants'[8] but it is generally accepted that some natural motion of the sea is necessary to disperse the oil effectively. As a result, aircraft spraying of dispersants has been introduced in which the concentrated product is sprayed undiluted but at the same absolute rate per unit area. New ship systems for the application of undiluted concentrate have also been developed.

Assessment of Performance and Toxicity

The use of dispersants in waters around the United Kingdom is now governed by the Dumping at Sea Act 1974. MAFF administers a scheme for the licencing of approved products and issues either 'User' Licences for bodies which may apply dispersants or 'Manufacturer-User' Licences for dispersant producers. Approval is granted or refused on the basis of performance and safety tests carried out by WSL both in the laboratory and at sea. If these are successful, the product is passed to MAFF for assessment of its potential toxic properties to marine life.

Although the WSL and MAFF tests are accepted by some other countries, in certain areas of the world different tests, more suited to local conditions are prescribed by the authorities.

Laboratory Tests

(a) *Performance:* Formerly, WSL assessed the performance of dispersants in the laboratory by means of the 'swirling beaker test'.[9] In this test, crude oil is floated on the surface of salt water in a number of beakers. Measured amounts of the dispersants under test are added so that they can be compared with one another and with a standard. The beakers and their contents are then swirled at standard speeds to show how much agitation is required to disperse the oil within various periods of time.

More recently, WSL has adopted a test method based on that developed by Labofina, Belgium[10] which, it is believed, relates more closely to the performance of dispersants at sea. In this test, crude or fuel oil is mixed with a measured amount of the dispersant under test. The oil/dispersant mixture is agitated by placing it in 250 ml conical separating funnels which are rotated about a horizontal axis for one minute at 34 rpm. The rotation is achieved by placing the funnels in a frame driven by an electric motor (Fig. 8.2).

The emulsion-forming efficiency of the dispersant is determined by running off aliquots of the aqueous layer after a period of time and measuring the amount of oil dispersed in it. This is done by a colorimetric method against a standard.

Another laboratory test (often referred to as the Sunbury Tank Test), developed by the British Petroleum Company Limited has now been adopted by the South African Bureau of Standards.[11] In this test, a small amount of crude oil is introduced into the centre of a plastic ring which is floating on the surface of salt water contained in a rectangular tank (Fig. 8.3). Dispersant is added and wave motion is simulated by means of an oscillating plate. Again, the dispersant under test is compared with a standard for its efficiency at dispersing the oil after a given period of time.

Recently, Mackay and Nadeau[12] have suggested a test method which in many ways resembles the Sunbury Tank Test. However, the wave motion is achieved by blowing air over the surface of the water in the tank.

A further performance test employed by, amongst others, the United States Environmental Protection Agency[13] is one in which oil is applied to the surface of synthetic sea water contained in a cylindrical tank. The dispersant is then applied to the oil in a fine stream and a one minute contact time is allowed. Energy is im-

Fig. 8.2 Laboratory apparatus for assessment of the performance of dispersants

parted to the oil/dispersant mixture by hosing and agitating the mixture with a pressurized water stream. The contents of the tank are then allowed to circulate via an external pump, being withdrawn from the bottom of the tank and returned just below the surface of the liquid (Fig. 8.4).

Samples are withdrawn from the tank after 10 minutes and 2 hours circulation, the oil extracted with chloroform and the quantity of oil present in the samples determined colorimetrically. The results obtained are compared with those for a 'standard' dispersant.

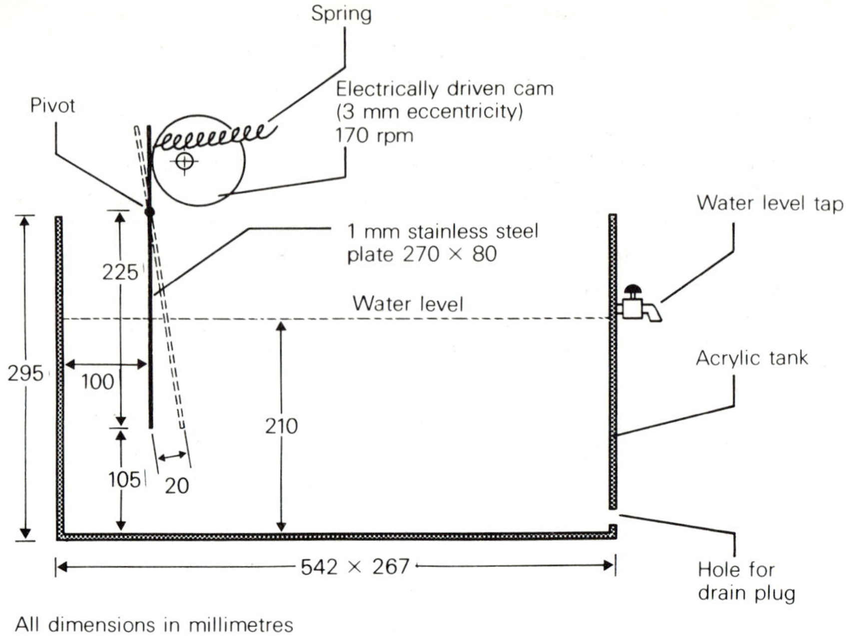

Fig. 8.3 South African Bureau of Standards oil-spill disperants test apparatus

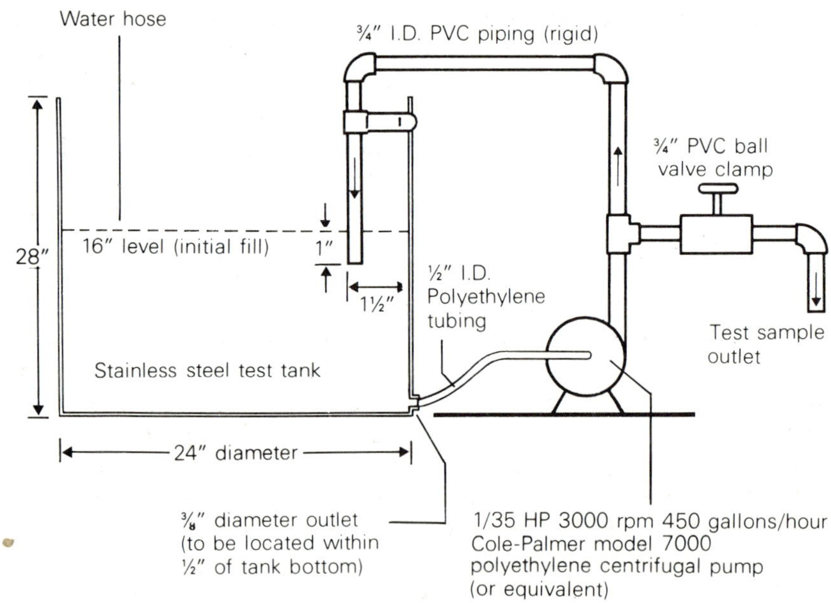

Fig. 8.4 United States Environmental Protection Agency oil-spill dispersant apparatus

(b) *Physical Properties:* WSL specify[14] a number of physical characteristics which a dispersant must exhibit and the use of certain chemicals in the product is prohibited. An abridged specification is as follows:

> A single phase liquid containing no solid or suspended matter.
> The product shall not be corrosive to its storage containers.
> The product shall not contain benzene, carbon tetrachloride or other chlorinated hydrocarbons, phenol, cresols, caustic alkali or free mineral acid.
> Flash point (Pensky Marten closed cup) 61°C min.
> Cloud point (IP method 219/67) −5°C max.
> Separation into two layers −10°C max.
> Viscosity at 0°C, 50 cS max. (250 cS max. for concentrates).
> Aromatics content of hydrocarbon solvent 3% max.
> Shelf life of product in sealed containers, 5 years min.

(c) *Toxicity:* Tests to assess the potential toxicity of products to marine life are normally carried out only on those dispersants which meet fully the performance and physical requirements.

Until fairly recently, toxicity testing was done under static conditions according to the method of Portmann and Connor[15] using the brown shrimp (*Crangon crangon*) and bullhead fish (*Agonus cataphractus*) as the test species. This test determines the 48 hour LC_{50} value of the dispersant, i.e. the concentration necessary to kill 50% of the test species within 48 hours under the specified test conditions. The 48 hour LC_{50} value of the first generation of (toxic) dispersants was of the order of 10–100 ppm whereas the second generation (low toxicity products) had values as high as 3300–10,000 ppm, both against the brown shrimp.

As dispersants of even lower toxicity have become available, the test concentrations necessary to derive 48 hour LC_{50} values have risen to levels where the dispersant no longer mixes adequately with the sea water in the test tanks but forms a thick floating emulsion or flocculate. The test animals are therefore liable to be exposed to concentrations of dispersant very different from those theoretically present, thus giving unreliable results.

New methods have been developed by MAFF to overcome this problem and they are described in their Fisheries Research Technical Report No. 39.[16] Two different test procedures are used in recognition of the different conditions of use of dispersants at sea and on beaches.

The sea test compares the toxicity of oil dispersed under standard conditions of mechanical agitation with that of the same amount of oil dispersed with an equal quantity of dispersant (or

10% dispersant in sea water in the case of concentrates) under the same conditions of mechanical agitation. The test is carried out in a perspex tank fitted with a stirrer which is covered by a plastic mesh to protect the test animals as shown in Fig. 8.5. Twenty animals are used for each test, one test with oil alone at a concentration of 1000 ppm and the other with oil at 1000 ppm plus dispersant at 1000 ppm. The exposure time is 100 minutes after which the animals are carefully transferred to clean sea water and, after 24 hours, each tank is inspected for dead animals and the percentage mortality calculated.

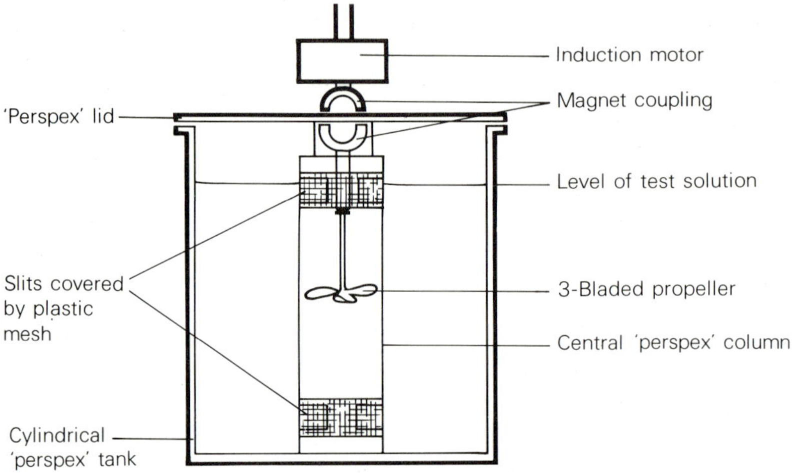

Fig. 8.5 Diagram of sea test tank for assessing toxicity of oil slick dispersants

The acceptable toxicity limit for the purposes of Licencing under the Dumping at Sea Act 1974 is that the toxicity of the oil plus dispersant should not be significantly greater than that of the oil alone under standard test conditions. This being so, it is considered that the use of that particular dispersant should not increase significantly the damage to the marine environment that is likely to result from exposure to the untreated oil dispersed by natural physical means.

The beach test is designed to assess the effect of dispersant spraying operations on a typical inter-tidal grazing organism, the common limpet (*Patella vulgata* L.), the actual test procedure being based on earlier work by Dicks.[17] Dispersants for use on beaches are approved on the basis that they should be no more toxic to the test organism than oil itself, on the premise that the use of dispersant on organisms that are unoiled be no more damaging than the effect of oil in the absence of spraying. The damage likely to be caused both by the oil and the dispersant is assessed by determining the number of limpets which drop off, under the standard conditions of the test, from perspex plates to which they have previously been allowed to attach themselves.

Sea Trials

There is, at the present time, only limited correlation between even the best laboratory performance tests and actual efficiency of dispersants at sea. It is therefore imperative that no dispersant approval should be given solely on the basis of laboratory tests. Thus, WSL carry out performance efficiency trials at sea which consist of what may be described as a 'titration' method of assessment. The basic idea of the test is to spray dispersant at a constant rate onto oil which can be sprayed at a changing rate. The rate at which oil is sprayed is increased until a point is reached at which the dispersant just fails to cope with the oil. Trials should be carried out only under relatively calm conditions.

For this, two boats are necessary. The (normally) larger work boat will be a tug or similar vessel capable of maintaining a steady speed of about 8 knots. The second boat can be any vessel capable of carrying the observers plus some means of spraying dispersant and agitating the water after spraying.

Dispersant is sprayed at a constant rate (4 gallons (18 litres) per minute) from a spray boom which is mounted about 1 m behind another nozzle which sprays oil at a variable rate.

The dispersant is turned on first followed by oil at 2 gallons per minute. If the observers in the second boat radio that the oil is adequately dispersed at this rate, then the oil rate is increased until the observers indicate that undispersed oil is visible as small droplets rising to the surface and coalescing to form larger patches of oil. The oil/dispersant mixture is agitated by a breaker board agitator of standard WSL design towed about 4 m behind the dispersant spray boom. When a test fails to disperse all the oil, this is dealt with by the spraying equipment in the observer boat.

Full details of the test and the equipment used are given in a WSL report[18] and the layout is shown schematically in Fig. 8.6.

A standard dispersant must be included with every batch of products tested. Owing to variations in weather and other conditions, the results of separate batches of tests carried out on different occasions cannot be quantitatively compared. Thus, dispersants can be ranked in order of efficiency but their effectiveness cannot be assessed in absolute terms.

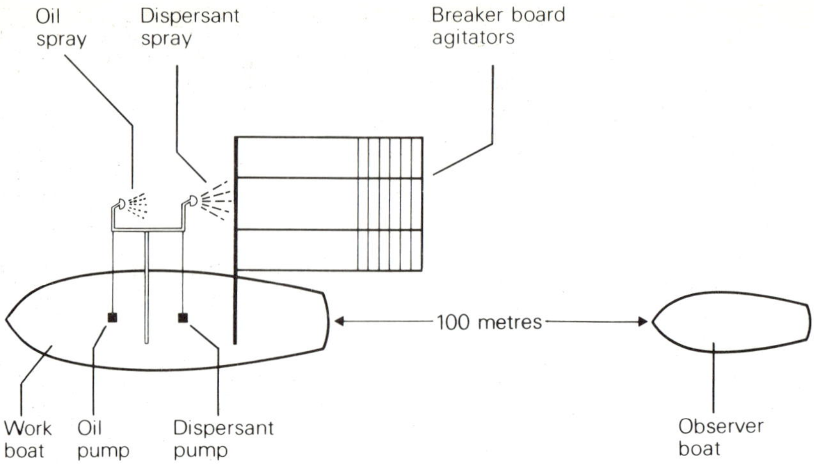

Fig. 8.6 Sketch of the Warren Springs Laboratory trial for oil dispersants

Spraying Equipment

(a) *Ship-borne*

Dispersants will not perform at maximum efficiency unless suitable equipment is used to apply them directly to the floating oil and to provide adequate agitation for the oil/dispersant/sea water mixture. The normal method of applying dispersant to the oil is by means of spray booms fitted to ships and the size of the actual spray nozzles should be chosen such that the dispersant droplet size resembles moderate rainfall. A certain amount of mixing energy for the treated oil is provided by the natural motion of the sea but this, even when augmented by high pressure water jets or the use of ship's propellers is often insufficient to ensure optimum efficiency of the dispersant and will only treat a small area at a time. This led to an investigation into the problem by Warren Spring Laboratory and, subsequently, a highly effective surface agitator which resembles a 5-bar gate was developed for towing behind the spraying vessel (Fig. 8.7).

The Warren Spring Laboratory developed a complete, fully demountable dispersant spraying kit[19] which can be erected easily on most ocean going tugs and larger fishing vessels without any welding or permanent fixing of any part of the equipment: it is easily assembled at sea. The dispersant is sprayed by means of the self-priming pump on to the sea from booms, mounted on both sides of the vessel, each fitted with three spray nozzles. These nozzles are specially selected so that a uniform pattern of droplets similar to a moderate rainfall is produced, thus ensuring that this pattern is only slightly affected by windy conditions. In order to

Fig. 8.7 Warren Spring Laboratory surface agitator in use

maintain this spray pattern, the dispersant pump rate must be constant at all times and the quantity of dispersant applied to the oil is varied by altering the speed of the work boat. The spray booms tow the surface agitators (three each side) which effectively mix the treated oil with the top 25 cm or so of sea water. A suitable vessel for use with this equipment is one capable of speeds of between 5 and 10 knots in moderate sea conditions and which can carry adequate supplies of dispersant. This is sprayed at the rate of 20 gallons (90 litres) per minute the optimum speed of the work boat normally being about 8 knots (Fig. 8.8).

The spraying equipment described above was developed originally for the application of hydrocarbon solvent based dispersants. With the advent of products of the concentrate type, modifications were made consisting essentially of diverting the main inlet to the pump so that sea water is drawn in over the side of the work boat and adding a small injector pump for the concentrated dispersant. This is shown schematically in Figs 8.9 and 8.10.

The inlet for the hydrocarbon solvent based dispersant in the conventional system becomes the inlet for sea water when the system is changed to spray concentrate type dispersant. This equipment, which can be used for the spraying of either neat hydrocarbon solvent based dispersant or dilute concentrate dispersant is available in kit form in three sizes from the suppliers whose name and addresses are given in Appendix 3.

Fig. 8.8 Ocean-going tug spraying dispersants with WSL deep sea spraying equipment showing effect of agitators

The three sizes in which the kits are available are as follows:

1. The major set for ocean going tugs and similar sized vessels of up to about 45 m overall.
2. The midi set for fishing boats of up to about 16 m long and for work launches which are often employed in coastal waters and off-shore unexposed areas.
3. The mini set for small work boats up to about 9 m long with minimum engine power of 15 horse power as used in harbours, estuaries and in-shore sheltered waters (Fig. 8.11). This mini set also incorporates beach cleaning attachments in the form of two hand held spray lances.

Aerial spraying has indicated that agitation of the oil/dispersant/sea water mixture by the WSL surface breaker boards which limit the speed of a spraying vessel to 5–10 knots may not be necessary if undiluted concentrate dispersant is used.[20,21] This has led to the

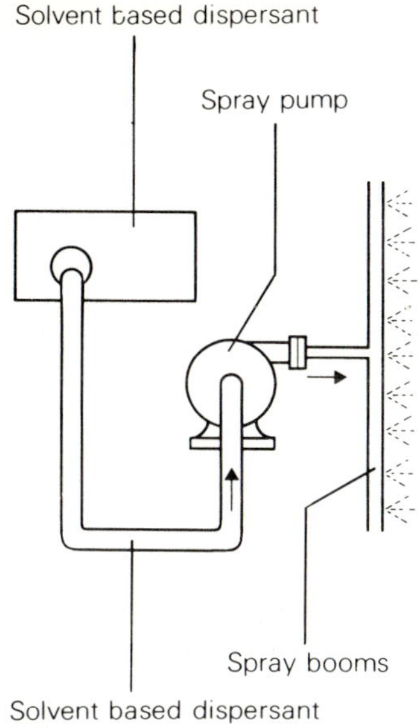

Fig. 8.9 System for spraying conventional hydrocarbon solvent-based dispersants

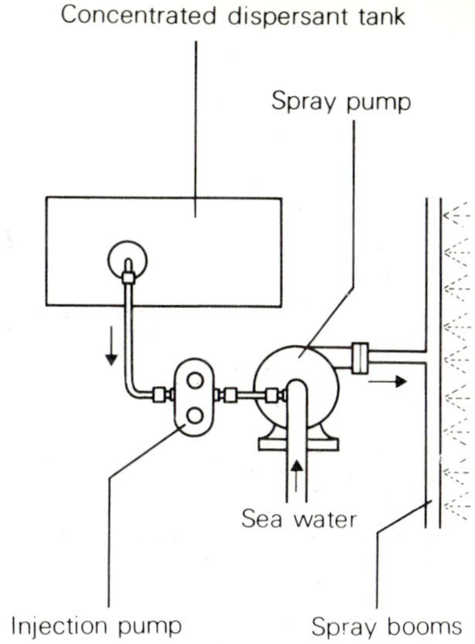

Fig. 8.10 System for spraying concentrate-type dispersants

Fig. 8.11 Small boat fitted with WSL inshore dispersant spraying equipment

development (by the manufacturers of the standard WSL equipment) of a more sophisticated system known as 'Widespray', with four spray jets each side of the vessel. This is for the larger ship with speeds of up to 12 or 15 knots and is for the spraying of undiluted concentrate dispersants. A ship fitted with 'Widespray' equipment is able to treat approximately 4 times the area in a given time compared with a similar vessel fitted with the standard WSL spray booms.

Another equipment manufacturer (Appendix 4) has developed a ship system for spraying undiluted concentrate dispersants using 'raindrop' nozzles producing droplets of 1000 μm size—claimed to be the optimum. The equipment is suitable for mounting on large tugs and other ocean going vessels. According to the manufacturers, it can be used in very rough sea conditions 'including gales of up to force 9'.

(b) *Air-borne*

Ships fitted with the WSL 'major' spraying equipment can, when correctly operated, disperse only about 15 tons of oil per hour per unit. Thus, if the spill is large, either a long time will elapse before the oil is treated or many spraying vessels must be used. Unless dispersion can be carried out quickly there is more chance of the oil being converted into 'mousse' or of coming ashore. The necessity for speed of treatment led to the use of aircraft to apply undiluted concentrate type dispersant. The high 'encounter rate' by aircraft means that large quantities of oil can often be treated before the formation of the less amenable 'mousse' occurs.

An aircraft travelling at 200 knots can potentially treat 20 times the amount of oil treated by a ship travelling at 10 knots but there is, of course, a payload limitation. Cormack[22] has compared the use of tugs at a hire charge of £2000 per day with an aircraft, having a 10 ton payload, at £500 per hour hire charge and a retainer fee of £100,000 per annum. The results are shown in Table 8.2.

TABLE 8.2 Relative Costs of using ships and aircraft for spraying Dispersants.

		Aircraft				Ships		
Spill size kilotons	*Cost of dispersant £'000*	*Operating time radius 100 miles hours*	*Total cost £'000*	*Cost/ton of oil dispersed £*	*Cost/ton with no retainer £*	*Operating time tug/days*	*Total cost £'000*	*Cost/ton of oil dispersed £*
2	75	10	180	90	40	20	115	575
6	225	30	340	64	40	60	345	575
10	375	50	500	50	40	100	575	575

He points out that, from the table, it is clear that:

1. Operating time for the aircraft is less than for a tug, confirming the advantage of aircraft to treat oil quickly.
2. If annual retainer is neglected for aircraft it is always cheaper than the tug.
3. Overall costs are swamped by dispersant costs, so the more effective use of aircraft to both seek out and then treat the oil makes most effective use of the dispersant.
4. Even with the retainer, if more than one 5000 ton spill occurs per year, aircraft cost would be less than using tugs.

To date most operations and experiments have been carried out using small single engined crop-spraying aircraft. These are usually available, and are already fitted with a tank and spray gear which can easily be modified to provide the necessary large droplets. However they have a limited pay-load and operating range. Nicholls and Parker[23] have examined this situation and calculated the time needed to disperse a 4000 ton oil spill using different types of aircraft. Table 8.3 from their paper, indicates that the large multi-engined aircraft, as used by Cormack, is more appropriate and would ensure a safe and cost effective operation. It is emphasised that during an actual operation the larger aircraft was found to be more difficult to manoeuvre when treating small slicks at low altitudes. The authors believe that, in the future, aircraft will be used more and more but no single aircraft, fixed wing or rotary wing, can be judged suitable for all oil spill situations and so several types may be necessary to provide for all circumstances. It remains true that for small, infrequent spills provision for equipment which can be mounted on any available vessel is the best precaution to take. Figure 8.12 shows a Piper Aztec equipped to spray dispersants.

TABLE 8.3 Comparison of oil treatment rates for different aircraft.

Aircraft	*Dispersant Payload (tons)*	*Time to treat 4000 tons of oil (h)*
Piper Pawnee	0.5	200
Piper Aztec	1.0	100
Douglas DC6	10.0	10

Fig. 8.12 Piper Aztec fitted with dispersant spraying equipment

Rate of Usage

The ratio of oil: dispersant necessary for effective dispersal of the oil will vary considerably from spill to spill and will depend very much on the state and type of oil being treated. One part of dispersant (at use concentration) may treat up to 10 times its volume of freshly spilled light oil in thin films. However, application rates as high as one part of dispersant to one part of oil may be necessary for heavy oils and mousses. An average rate is about one part of dispersant (at use concentration) to three parts of oil or 1:30 if neat concentrate is used.

Ecological Factors

There are normally few ecological problems associated with the use of dispersants in off-shore situations. The latest low toxicity approved dispersants are unlikely to reach toxic concentrations, as dilution in open sea conditions is very rapid. However, in inshore situations, there is normally less rapid dilution of the dispersed oil. In these situations, shell fisheries, fish nursery areas and other sensitive resources may be at risk and it is therefore imperative that local fisheries authorities be consulted before spraying operations are started.

Methods of Application

It is absolutely essential that, to achieve good results, oil slick dispersants be applied as recommended by the manufacturers. Much dispersant has, in the past, been wasted by incorrect usage and this has, on occasions, given dispersants a bad name. If a product is incorrectly applied, a quantity highly in excess of that normally required has to be used and, especially in estuarine situations, this can result in a build-up of higher concentrations of dispersant in the water than are necessary. Dispersants based on hydrocarbon solvents must, in general, be applied neat to the oil, whereas the concentrate type must be diluted with sea water prior to application when the standard ship spraying system is used.

Spraying Techniques

It is important to note that full protective clothing (including goggles and face mask) should be worn when spraying dispersants. They should also be applied strictly according to the manufacturers instructions and under trained supervision. Hydrocarbon solvent type dispersants must never be mixed with concentrates.

The spraying techniques vary according to location and type of oil spilled. If the slick is out on the open sea, well away from the coast, it is probably best to leave it to disperse naturally.

For a larger slick which is on the open sea but relatively close to the coast, if the oil is amenable to dispersant treatment, it should be sprayed continuously along its outer edge working towards the centre to avoid fragmentation. Often, oil spilled at sea is fragmented naturally into 'windrows' in which case they should be sprayed up (and down) working into the wind. If the oil is very near to land, spray along its landward edge and parallel to the land. If two or more spraying vessels are available, it is best to operate them at a distance so that the wake from one does not interfere with the spraying operations of the other. The speed of the work boats is important and no advantage is to be gained from either very slow or very fast steaming. The optimum speed in any given situation will depend upon the thickness of the oil film and its viscosity but, as a general rule, when the conventional WSL equipment is used, speeds below 4 knots produce insufficient agitation and speeds above 10 knots result in the breaker boards rising over the surface of the treated oil.

An aerial survey of the spill is highly desirable to determine the extent of the pollution and to assess where the vessel should start spraying. As stated earlier, the spraying of undiluted concentrate dispersants from aircraft is now becoming an accepted technique.[24] One advantage of aerial spraying is that both the survey and the spraying can be carried out more or less simultaneously.

In addition, the 'encounter rate' can be increased thus reducing the time between spillage and treatment and the possibility of weathering and mousse formation. Recent work has indicated that mousses may form very rapidly after a spill, the water content possibly reaching 70–80% within 24 h.[25] Dispersant treatment should thus be effected as soon as possible following the spill.

Oil spillage can occur in ports, harbours and other sheltered waters and here the quantity of oil spilled will normally determine the best method of clearance. Dispersants alone should not be used to deal with very large quantities of oil in confined spaces and, in such instances, mechanical oil recovery devices or absorbents should be employed to remove the bulk of the oil followed by treatment of the remaining oil with dispersants. For smaller spillages, dispersants alone will be the quickest and most effective method of treatment. Oil floating well away from quaysides can be treated with dispersant using the WSL in-shore (mini) spraying equipment: oil close to or actually on harbour walls may be treated either from the water or the land with the WSL hand lances, as used for spraying beaches. For vertical surfaces and under piers, the use of breaker boards is not feasible and agitation by means of high pressure water jets played as near tangentially as possible to the surface of the treated oil is recommended. It must be emphasised that dispersion by means of high pressure water is an expedient which should only be used in cramped conditions since it is not as efficient in dispersing the oil as agitation provided by breaker boards.

Oil on Inland Water

The use of dispersants is not recommended since the volume of water is normally insufficient. Relatively high concentrations of oil and dispersant are produced in the water and there is no natural agitation as is present in the sea. Therefore, mechanical recovery or the use of absorbents is recommended.

Conclusion

Dispersants have, over a period of time, proved to be an effective and practical method of dealing with large and small oil spillages both at sea, in-shore and on beaches. It is important that they be used according to the manufacturers instructions and with suitable and efficient spraying equipment such as that developed by the Warren Spring Laboratory. It is not sufficient simply to have supplies of dispersant and equipment available at the right place at the right time: trained personnel must also be available. It must also be borne in mind that it is much easier to disperse oil whilst it is still at sea and every effort should be made to do this before it comes ashore. This will be facilitated by the advent of the application of undilated concentrate type dispersants from aircraft.

References

1. WARDLEY-SMITH, J., 'UK.Min.Tech. work on oil pollution', EPA, API, USCG, Conference on prevention and control of Oil Pollution, New York, 1969.
2. WARDLEY-SMITH, J., and SHELTON R. G., 'Disposal of oil spills from water surfaces', Proceedings 8th World Petroleum Congress, Moscow 1971.
3. TADEHA, H. J., *The Shell Oil Sinker Method*, Koninklijke/Shell Exploratie en Produktie Laboratorium, Rijswijk, The Netherlands, 1970.
4. CANEVARI, G. P., 'Development of the Next Generation Chemical Dispersants', Proceedings of Joint Conference on Prevention and Control of Oil Spills, API, Washington DC, 1973.
5. MARTINELLI, F. N. and CORMACK, D., *Investigation of the Effect of Oil Viscosity and Water in Oil Emulsion Formation on Dispersant Efficiency*, Report LR 313 (OP), Warren Spring Laboratory, Stevenage, Herts, UK, 1979.
6. MARTINELLI, F. N. and LYNCH, B. W. J., *Factors Affecting the Efficiency of Dispersants*, Report LR 363 (OP), Warren Spring Laboratory, Stevenage, Herts, UK, 1980.
7. SIMPSON, A. C., '*Torrey Canyon*' *Disaster and Fisheries*, MAFF Laboratory Leaflet (New Series) No. 18, February 1968.

8. CANEVARI, G. P., 'A Review of the Utility of Self Mixing Dispersants in Recent Years', Proceedings of Conference on Prevention and Control of Oil Pollution, API, San Francisco, 1975.
9. *The Swirling Beaker Test for Oil Slick Dispersants*, Warren Spring Laboratory, Stevenage, Herts, UK.
10. VON EMMANUEL, C. C. and LEPAIN, A. O. *Erdoel Kohle, Erdgas, Petrochemie Brennst-Chem.* **10**, 646–651 (1971).
11. South African Bureau of Standards, *Standard Specification for Oil Spill Dispersant (Liquid-Solvent Type)*, Ref. SABS 1025–1975.
12. MACKAY, D., NADEAU, J. S. and NG, C., *A Small Scale Laboratory Dispersant Effectiveness Test*, Report of Department of Chemical Engineering and Applied Chemistry, University of Toronto, November 1977.
13. United States Military Specification, *Solvent Emulsifier, Oil Slick (Test 25)*, MIL-S-22864A Ships, 24 February 1969.
14. CORMACK, D., LYNCH, B. and SMITH, J., *Dispersants for Oil Spill Clean-up Operations at Sea, on Coastal Waters and Beaches*, Report LR 316 (Op), Warren Spring Laboratory, Stevenage, Herts, UK, 1979.
15. PORTMANN and CONNOR, *Marine Biol.* **1**(4), 322–329 (1967).
16. BLACKMAN, R. A. A., FRANKLIN, F. L., NORTON, M.G. and WILSON, K. W., *New Procedures for the Toxicity Testing of Oil Slick Dispersants*, Report No. 39, Ministry of Agriculture, Fisheries and Food, Directorate of Fisheries Research, Lowestoft, Suffolk, UK, 1977.
17. DICKS, B., 'Some effects of Kuwait crude oil on the limpet (*Patella vulgata L.*)', *Journal of Environmental Pollution* **5**, 219–229 (1973).
18. *A Method of Testing Oil Dispersant Chemicals at Sea*, Report LR 152 (ES), Warren Spring Laboratory, Stevenage, Herts, UK, 1971.
19. *Development of the WSL Dispersant Spraying Equipment*, Report LR 151 (ES), Warren Spring Laboratory, Stevenage, Herts, UK, 1971.
20. CORMACK, D. and PARKER, H., 'The Use of Aircraft for the Clearance of Oil Spills at Sea', Proceedings USCG, API, EPA Oil Spill Conference, Los Angeles, USA, 1979, p. 469.
21. CORMACK, D and NICHOLS, J. A., *Feasibility Study of Aerial Application of Oil Dispersant Concentrates for Oil Spill Clearance*, Report LR 257 (OP), Warren Spring Laboratory, Stevenage, Herts, UK, 1977.
22. CORMACK, D., 'Relative, Effectiveness of Current Techniques for Oil Spill Clearance', *Petroleum and the Marine Environment*, Graham & Trotman, London, 1981, pp. 135–147.
23. NICHOLS, J. A and PARKER, H. D., 'Aircraft application of oil slick dispersants at sea', International Agricultural Aviation Conference, Turin, 1980.
24. NICHOLS, J. A. and WHITE, I. C., 'Aerial Application of Dispersants in Bantry Bay following the Betelgeuse Incident 1979', *Marine Pollution Bulletin* **10**, 193 (1979).

25. CORMACK, D., NICHOLS, J. A. and LYNCH, B., *Investigation of Factors affecting the Fate of North Sea Oil Discharged at Sea, Part I Ekofisk Crude Oil July 1975–Feb 1978*, Report LR 273 (OP), Warren Spring Laboratory, Stevenage, Herts, UK, 1978.

Part Three

Treatment of Oil on the Land

9

Sea and freshwater spills

Oil on the shore

There are many reasons why oil can reach the shore. It may have come from a large-scale spill or an accident to a tanker, and, despite all efforts to disperse or collect the oil, it is almost inevitable that some will come ashore. Sometimes slicks are observed at sea which are not apparently the result of an accident and neither a ship nor a tanker is visible. Floating oil from an unidentified source is usually not dealt with at sea, as experience has shown that reports of slicks of this kind are unreliable and often exaggerate the quantity. However, if such oil is deposited by wind and tide onto the shore it may be necessary to remove it. The commonest sort of oil found on the shore is the lumps of oil or 'tar', particularly frequent on coasts exposed to a prevailing wind. Under some circumstances beach cleaning may be needed as a result of this kind of pollution.

Removing oil from the land is seldom an easy matter and no single method or device can be recommended. The situation is complicated by the different physical characteristics of the oils which are contaminating and the many different types of shore line which can be anything between soft mud and hard rock.

It was emphasised in Chapter 4, on the inter-tidal zone, that oil pollution which is heavy enough to warrant removal from the shore line is unlikely to cause much environmental damage, while careless or unskilled attempts at removal may easily do much more harm than the oil itself. There is little doubt that in many cases the environment is likely to suffer most if an attempt is made to remove the oil. As described earlier, oil slowly disappears through natural causes. If it has been deposited on the shore, where it will be under water at high tide, then degradation is more rapid and also some of the oil will be removed by successive tides. Sometimes oil is carried ashore by a Spring tide and so may very seldom be covered by the sea. In this case the oil could remain a long time. For example water-in-oil emulsion from the *Torrey Canyon*, thickly deposited in a private cove on the Channel Island of Guernsey, had been reduced in one year to a dark paint-like layer which was dry and hard on some of the rocks, whereas oil from the *Esso Essen* cast high on the shore of Cape Point, South Africa, was found to be still 'oily', albeit in a very small patch, nearly three years later.

Before any cleaning is attempted two questions must be considered:

1. Why is cleaning or oil removal required?
2. Will the process damage the environment?

If the oil is on a beach used by holiday makers or residents and the beach is actually in use (there is little purpose in cleaning a bathing beach if it will not be used for many weeks) then clearly the clean-up must be quick and thorough. If the beach is not in use then the process should be delayed as long as possible and only done just before the area returns to use. By this means, natural clean-up will have been used to the maximum and any further accumulation of oil, after the initial landing, will also be removed with the single operation.

Where the coastline has a low amenity value, or is seldom used, then clean-up should not be carried out. Oil deposited on the tide line will be refloated by a subsequent higher tide if this time there is an off shore wind, and if sea conditions are very different, then it is possible that the oil will be completely taken away to sea.

This floating oil could be redeposited on a shore line which has a higher amenity value or one which could be damaged by the oil. Thus the decision to remove the oil or to leave it on the shore is not an easy one, and in some cases oil may have to be removed from a low amenity area.

Even in a vacation area and during the season there are many types of coast, for example, rocky shores or salt marshes, which are only occasionally used by people and where cleaning is not of first priority. In any area which oil might pollute, the whole coast line should be examined in detail by scientists who are familiar with the district, by the Nature Conservancy (in UK) or the Fish & Wild Life Organisation to determine, in collaboration with organisations concerned with the holiday makers (Tourist Board or Hotel Keepers Association), which coast line must be cleaned and which need not. All this must be done before an oil spill occurs—it cannot be left until the oil is on the shore. In the case of areas to be cleaned, the precise type of cleaning permitted, and the extent of the operation, should have been decided in advance and included in the maps of the contingency plan (see Chapter 10).

It must be emphasised that the movement of people over an area in trying to clean it up is sometimes enough to cause damage. For example, vehicles and people crossing sand dunes to reach the scene of pollution, can so damage the plant cover and expose the sand that it can be blown away by the wind, giving permanent or very long-term damage. Shore marine life is highly sensitive to changes in environment so that hosing down a rocky shore can dislodge certain plants and animals, while others may be killed by the reduction in salinity of the rock pools in which they live.

The possible short-term reduction in amenity due to the presence of oil must be balanced against the possible environmental damage, short or long term or even permanent, which can result from the misplaced enthusiasm of, often amateur, cleaning squads.

Methods employed for cleaning oil from beaches

The most obvious way of beach cleaning is to remove the oil physically. This is far from easy on many beaches as the oil gets mixed with, and coats, the beach material. Consequently, a number of other methods have been suggested; these include burning the oil, covering it up with an absorbent powder or washing it away using a dispersant chemical. Each of these methods could be of value in particular circumstances, and they are considered below.

Mechanical removal

From all points of view, mechanical removal, even if only some of the oil can be taken away, is the most satisfactory method. There is then no danger that the oil can get into the sea or remain to be a nuisance, and it reduces the quantity of dispersant chemical needed to complete the cleaning.

A sand beach, firm enough to bear vehicles and with access from a road, can be cleaned by removing the minimum amount of oily sand which can be collected by mechanical graders, or bulldozers, and loaded into lorries by fore loaders. To reduce cost this process must be carried out with great care to ensure that only oily sand is removed and that some of the dirty material is not spread out over a previously clean beach. The sand on many beaches is replenished by the action of tide and waves, but in some places the reverse is the case and sand is lost. Then, even more care must be taken to ensure the removal of a minimum quantity, otherwise it may be necessary to bring in clean sand as a replacement—usually a costly undertaking.

Viscous oil or lumps of oil deposited along the high tide line on a smooth sand beach can be collected by the use of a horticultural rake of fine spring loaded tines, of the type which is used on sports grounds to improve the turf. Alternatively, an agricultural hay rake can be used if the tines are interwoven with hay or straw to collect the finer lumps. Either of these devices can be used to collect the oil into rows or heaps for manual loading into barrows or trucks for removal. Disposal of the collected oily material is discussed at the end of this chapter.

Many countries have developed special machines to collect tarry lumps and other trash from sandy, amenity beaches. Typical examples of these machines have been developed in the United Kingdom particularly for use on the beaches of the South West counties of the United Kingdom. The devices are only suitable for use on dry and wet sand beaches. They are not suitable for use on shingle or pebble beaches.

The machines are quite different in concept and design. One is wheeled and intended to be hauled by a tractor, while the other is mounted on a tractor to make a complete machine.

The wheeled device for the collection of tarry oil lumps and litter is mounted behind and towed by a tractor, (see Fig. 9.1). It has an integral petrol driven engine which supplies the necessary power to drive the mechanisms. Basically, the device consists of a sand cutting blade, a bar conveyor, an inclined vibrating bar screen, a rear mounted refuse container and a rear mounted sand levelling bar. In operation the cutting blade is set at the required depth into the sand; this allows the sand, tarry oil lumps and litter to be fed into the device when being towed. The petrol engine drives the bar conveyor and the vibrating bar screen. The bar conveyor carries the sand, oil and litter over the vibrating bar screen which allows the sand to gravitate through while the tarry oil lumps and litter are carried rearwards and upwards to the rear mounted refuse container. The rear mounted sand levelling bar ensures that any undulations in the beach caused by the device are removed.

Two bar screen sizes are available for use with the machine 0.55 inches (14 mm) and 0.22 inches (5.5 mm). The manufacturer recommends the device being towed by a wheeled tractor of 25 hp minimum and equipped with high flotation tyres.

Fig. 9.1 The Beemer beach cleaner or tarry-lump collector

The other device for the collection of tarry oil lumps and litter is mounted behind a standard Massey-Ferguson tractor (Fig. 9.2). It consists of a rotating conical screen with interchangeable screens of mesh sizes 12.5 mm (½ inch), 19 mm (¾ inch) and 25 mm (1 inch). Two mouldboards are fitted to the underside of the tractor which have an entry opening of 122 cm (48 inches).

In operation, the tractor moves forward with the rotating conical screen set a few inches into the sand surface. Sand, tarry oil lumps and litter pass through the mouldboard guides and into the rotating screen. The sand passes out through the screen and the tarry oil lumps and litter are carried rearwards and upwards by the two auger scrolls and deposited into the rear-mounted refuse container. A rear-mounted level bar ensures that any undulations in the beach caused by the device are levelled out.

A front loader is a necessary part of the equipment as it is required to counterbalance the rear-mounted device. It is also required for lifting into disposal vehicles tarry oil lumps and litter that the device has collected.

In trials, carried out by the Warren Spring Laboratory, both machines were reported to work well in dry sand; the device collected the tarry oil lumps from the surface of the beach and those buried in the sand. It also collected tarry oil lumps mixed

Fig. 9.2 Massey-Ferguson machine for recovering tar-balls and trash from a sand beach

with general beach litter. On wet sand, the small mesh screens tended to clog but the large mesh screens worked satisfactorily.[1] It is in use in a number of places in England.

It would be a considerable advantage, particularly for a sand beach which tends to lose its sand if, rather than removing the oily material, it could be cleaned and then replaced on the beach. One way of doing this is to use a dispersant. This method (described later) does not remove the oil from the marine environment, nor does it recover the oil, and in some countries the method cannot be used either on ecological or legal grounds, therefore machines have been developed to clean fine beach material and return it to the shore.

One machine, developed in 1967 in Germany, burned off the oil.[2] The oily sand was collected, using graders and fore loaders, and conveyed to the plant which was transportable so that it could be moved to a convenient place on the shore above high tide line. As there is not usually enough oil in the beach material both to dry it out and to maintain combustion, the material was fed into a rotating furnace which was heated by oil. Draught was carefully controlled to avoid smoke pollution. The beach material after treatment was dry and clean but rather dark in colour apparently due to the adhesion of some carbon. It could treat 12 tonnes an hour. One plant was built but it is understood that the operating costs were high.

A simpler machine to burn the oil contained in sand was developed in Canada.[3] This can be manufactured without the need for sophisticated engineering. Basically it is an oscillating tube furnace made by welding five 45 gallon oil drums end to end. A double stack chimney is formed from two columns, each of seven 45 gallon drums welded end to end with their ends removed. The kiln is oscillated by a crank arm which replaces one rear wheel of an ordinary farm tractor. Figure 9.3 shows the complete machine.

Fig. 9.3 Incinerator in use. Note push – pull rod to tractor on left

Another machine, developed in the USA under an EPA contract,[4] also treated and collected from the beach, but, in this case water-washing in a type of bubble flotation system worked the sand clean of oil and returned it wet, to be replaced on the beach. The machine was completed and demonstrated but it is not known if it was ever used.

The problem of removing oil from a sand beach is not easy, but that of removing it from a shingle or pebble beach is much more difficult. If the oil is thin, then it can run down between the individual stones, while a more viscous oil can coat the stones very heavily and run into, or be deposited in, the interstices between them. Removal of beach material with fore loaders is almost the only mechanical solution to the problem.

Removal of lumps of thick oil or 'tar' is also very difficult because the oil can resemble the stones in shape and size (Fig. 9.4). In some popular holiday resorts on the south coast of England which have shingle beaches, the lumps are picked up by hand using men traversing the shore line abreast; they also carry out normal scavenging, picking up bottles, cans and general debris dropped by visitors or brought in by the tide. Attempts have been made to develop machines which can pick up tar balls from the surface of a shingle beach. In general these have not been satisfactory. The machine may work but is heavy and has a limited application. The main defect was due to the beach rather than the machine. Shingle beaches of the type described are frequently highly mobile under wave action, and to prevent erosion they are usually divided by sea walls or groynes which effectively prevent the free movement of a large machine.

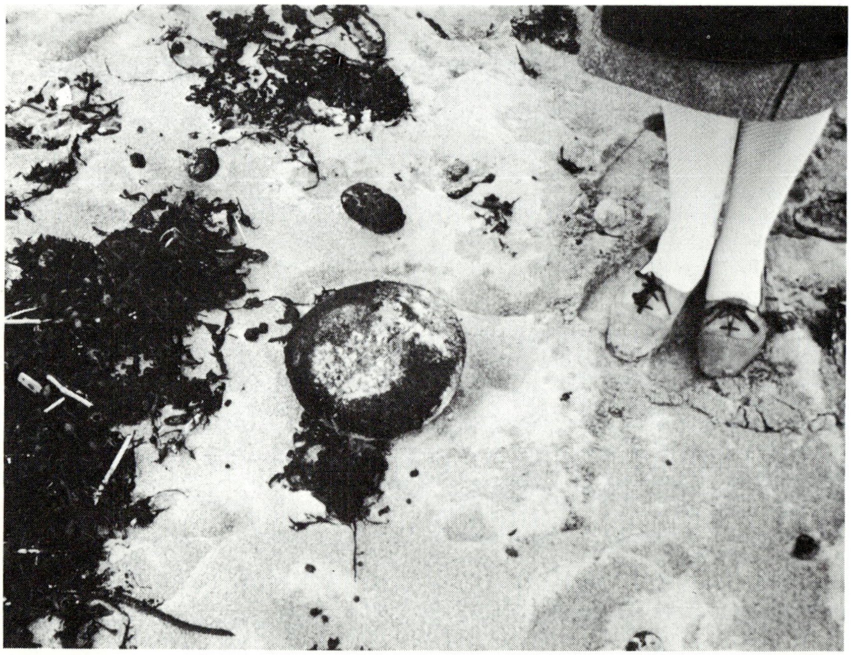

Fig. 9.4 Large 'tar ball' on a Cornish beach

Mechanical removal of oil on rocky shores can only be done by hand cleaning, aided perhaps by suction from gulley emptiers to remove oil from pools. However, as discussed in Chapter 4, even this degree of cleaning is undesirable.

Salt marshes, mud flats, mangrove swamps and similar areas which are either water-logged or partly covered at every tide cannot be cleaned mechanically. Where attempts have been made to clean these areas some destruction of the local environment has inevitably taken place.

Burning

The general problem of burning oil in and from a tanker is discussed in Chapter 7, and many of the difficulties described are also applicable to oil on the shore. There is an additional problem. Oil on the shore is often wet, either because the water is in droplets throughout the body of the oil, that is, on the way to the formation of a water-in-oil emulsion (mousse), or because the oil is mixed with seaweed and other wet detritus. It is very difficult even to ignite oil when it is on the shore, and impossible to get complete combustion. Lumps of oil mixed with flammable material such as driftwood and plastic, can be piled into heaps and set on fire. It will be found that the oil runs off materials like driftwood as it gets hot and runs into the beach material, and that as 'tar balls' melt the oil does the same. This oil does not burn so that the final result is most unsatisfactory; also, a large volume of unpleasant smelling smoke may be given off. If burning this sort of material is considered necessary, perhaps on an isolated beach difficult of access, then an improvised furnace such as a perforated 45 gallon drum with one end removed, will improve the air flow and catch melted tar. A pipe feeding in low pressure compressed air tangentially will help combustion.

Claims have been made that some of the ignition materials used to burn floating oil (see Chapter 7) can be used to burn oil on a beach, but although they help in lighting the oil, eventually the fire goes out, leaving a patch of ground heavily contaminated with unburnt heavy residue.

Use of absorbent material

Attempts have been made to use absorbent materials to remove oil from the ground. If the ground is smooth, as, for example a cemented jetty, an oleophyllic material, such as silicon treated sawdust, can be sprinkled over the oil and the whole lot brushed into heaps and removed. There are a number of substances which are sold on the market to clean garage and workshop floors and which are quite satisfactory. It can be seen that this method is impossible if the surface to be cleaned is made up of loose material like sand, or shingle. These absorbent materials therefore have a very limited use.

An alternative use of absorbent materials is to cover the oil with a powder to remove its 'oily' properties, and to leave the oil and powder to weather. A number of powders have been tried for this purpose, but none of them are really satisfactory unless very large quantities are used. Liquid oil is absorbed by the powder but tends to exude if pressure is applied, for example, by walking or sitting on it, and with lumps it is only possible to coat the upper surface.

Sterate-treated limestone dust, which is used for explosion control in coalmines, has been used for one particular oil spill in the United Kingdom[5] with reasonable success. The treatment makes the powder both waterproof and more ready to absorb oil. As it is a very fine powder, it is likely to be objectionable if used near houses in windy conditions. It has been suggested that treatment of this kind would be useful on isolated oil patches on marsh land to protect grazing wild fowl from becoming oiled. Quicklime can also be used for this purpose, its particular advantages being when the oily material is removed from the beach. This is discussed under 'disposal' later in the chapter.

The use of dispersants

It is normal practice to clean the floors of garages, workshops, and factories by applying a proprietary cleaning fluid, followed by hosing down with water and brushing the floor clean. The cleaning fluid usually consists of a powerful emulsifying agent dissolved in a hydrocarbon solvent. It is therefore logical to attempt to use the same or similar materials to clean oil-covered rocks and beaches in general. The whole question has been discussed in detail both in Chapter 4 dealing with fish and marine life and also in Chapter 8 dealing with the treatment of floating oil. It must, however, be reiterated that commercial factory dispersants are likely to be toxic to marine life, as indeed were those first used for beach cleaning. The currently available dispersants are more effective and much less toxic, the quantity necessary to kill test animals exposed for 48 h being a few parts per thousand, whereas the earlier materials had a toxicity of a few parts per million. Nevertheless, it will be appreciated that it does not take very much dispersant in a rock pool to produce a concentration which can be well above the toxic level, so that these materials should only be used with considerable caution.

There are three steps in dispersing oil. First, the oil must be treated with an appropriate amount of the dispersant chemical. Then, either time must be allowed for the dispersant chemical to become mixed with the oil or preferably some form of local agitation should be given, and last, the treated oil must be mixed vigorously with as much sea water as possible, both to produce a diluted dispersion and also to reduce the possible toxicity of the dispersing chemical by increasing its dilution.

This method should be used with considerable caution if there are any commercial fisheries, particularly shellfish, close to the shore being treated, or if there are nature reserves and, as discussed in the section on biodegradation, dispersants should not be used on the shores of fresh water lakes, rivers or canals, nor indeed anywhere where tidal movement is small or the sea area is confined as in a tideless bay or long estuary.

The precise amount of dispersant needed will depend on a number of circumstances—the type of oil, the efficiency of the dispersant, the type of beach, the temperature, all have an influence. As a rough guide, if the volume of dispersant is about 25% of the volume of the oil to be removed, this will produce reasonably effective cleaning. Even this figure may not be of much help, as it is very difficult to estimate the amount of oil actually spread in areas of varying thickness over a large stretch of beach. If possible, the oil should be sprayed with the dispersant about 30 minutes before the tide covers the site and the waves can agitate the dispersed oil. If the period is going to be much longer than an hour, then it is highly desirable that the treated oil is washed down with water, fresh or salt, applied from hoses such as fire hoses. This makes the cleaning very much better and also reduces still further the risk of damage to the inter-tidal flora and fauna.

The dispersants are frequently flammable, and being oil solvents, they can have a harmful effect on the skin by removing the natural oils, so care must be taken to see that the operators do not work for prolonged periods with either their skin wetted by the dispersant, or wearing clothing soaked in dispersant. PVC protective clothing should be worn, and if there is any danger of getting spray into the eyes, goggles also. After using the material, the hands and face of the operators, should be washed with soap and water, and lanolin or cream containing lanolin should be applied. The dispersant is a solvent for rubber and rubber materials; consequently rubber boots, rubber hoses and rubber washers in pumps and spraying equipment, will all be readily attacked. Gloves and hoses should be made of PVC or similar material, and washers for pumps should be of an oil and solvent resistant material.

Although it is possible to spray the dispersant using improvised sprayers such as knapsack sprayers of the garden or horticultural type, and even watering cans with fine roses, it is much better to use a properly designed lance with a spray nozzle fed by means of light hose and a motor driven pump. It was found (Fig. 9.5), that the operator can walk backwards and forwards, along as much as 30 m length of beach, dragging a light PVC hose behind him, while the pump and the supply of dispersant can be carried in a trailer or truck at the top of the beach. Pump Trailer Units which are self-contained with motor driven pump hoses and a tank of dispersant are available (Fig. 9.6). Some are arranged to use the newer concentrated dispersant drawing the correct amount of mixing water from the sea. If, after spraying, the tide is not going to cover the sprayed oil for some time, then it is essential that the treated oil is hosed back into the sea using fire hoses fed from suitable pumps, either using fresh or salt water.

Once more it must be emphasised that every effort should be made to remove as much oil as possible by mechanical means before attempting the final clean-up using a dispersant. If large quantities of oil are dispersed it will be found that the oil/dispersant mixture, rather than being mixed in with the sea and carried away, tends to sink down through the beach material. If this is fine sand then local 'quicksands' can be produced, and it will be found that the oil can be carried down through the sand until an

Fig. 9.5 Beach spraying using WSL mini pump

Fig. 9.6 Cooper Pegler beach trailer

impervious layer or the water interface is reached. At this level the oxygen content of the water is very much less and the bacterial degradation of the dispersed oil, if it takes place at all, will only be very slow, so that although the oil can apparently be lost from the beach surface it may well be uncovered, either by subsequent heavy wave motion or even by children or fishermen digging in the sand.

Drums of the dispersant should be stored, when not required, in well ventilated sheds, as if the drums were filled with kerosene. When the contents of the drums are transferred to other containers, care must be taken to see that nothing is spilled and particularly that it is not allowed to enter drains or run into water courses. Any spills should be soaked up with sawdust, peat, dry earth, etc. and then removed from the site. In this connection it is interesting to note that one of the last remaining pieces of evidence of the *Torrey Canyon* disaster which occurred in 1967[6] are patches of dead turf on the top of some of the cliffs.

Types of Coast

The treatment required to deal in the most economic, and at the same time environmentally satisfactory, manner depends, as has already been pointed out, on the nature of the shore. This section therefore is divided among the major shore types, including both arctic and tropical areas.

Salt marshes

This term is intended to include areas of grass and reeds which are covered only occasionally by high tides. They are usually intersected by small streams. The problem of contamination of these areas has been discussed in Chapter 3 and it will be appreciated that cleaning of any sort is particularly difficult, and almost anything that is done will add to the damage done by the oil. Collection by hand is usually the only satisfactory method, but in the late Summer, Autumn and Winter, use of a flame gun to burn dry vegetation and the adhering oil is reasonably successful. In general, these areas should be left uncleaned.

Mud flats

Mud flats are usually covered at every tide. If they become coated with oil, the next tide may well transfer the oil to another part of the coast where it may be of more nuisance to visitors and other users. Even so, it is very strongly suggested that no action should be taken. Any mechanical damage or alteration to the marsh profile can affect tidal movement over it with perhaps long-term effects.

Sand

The sandy beach is the type of beach which attracts the largest number of holidaymakers (Fig. 9.7). If, however, the beach is not likely to be used extensively for a period, for example, in winter, then initially nothing should be done. Cleaning up may be required immediately before a vacation season starts, but the delay will have reduced the quantity of oil to be removed. The best methods of dealing with the oil have been discussed earlier. Lighter oils may well sink into the sand, in which event the only solution is to plough up the area into deep furrows, spray with a suitable non-toxic dispersant and allow the sea to flatten the furrows and remove the oil. In any event, after any form of mechanical removal of the oil, light dispersant treatment followed by wave action or hosing down will usually be necessary.

Shingle

Shingle beaches (Fig. 9.8) heavily coated with thin oil are particu-

Fig. 9.7 Sandy beach

larly difficult to clean as the oil runs down through the shingle to a considerable depth. More viscous oils coat the shingle surface and are equally difficult to remove. Bulldozing away the top layer of contaminated material, followed by dispersant spraying and hosing down, or preferably wave action, is the only way to produce a clean surface. A shingle beach is to a great extent, self-cleaning by wave action, so that a delay in cleaning will give less oil to clean-up. Removal of tarry lumps may well have to be done by hand due to the presence of groynes used to stabilise the beach. These prevent any type of vehicle traversing the beach and so considerably hamper any large scale cleaning operation (Fig. 9.8). It will be found in many cases that a storm, by turning over the shingle, brings more tarry lumps to the surface.

Rocks and cliffs and man-made structures

Although flame guns, steam treatment and hot water washing have all, on occasion, been used to clean these areas, the use of any of these methods would virtually sterilise the rock surface and should be avoided. Heavy deposits of oil will have to be removed, probably by hand, followed by dispersant spraying and hosing down with powerful hoses, or wave action. If the surface is on a steep slope or vertical, the dispersant will run off before it has time to penetrate. Although, after a time, the oil will harden due to weathering, whilst it remains fairly innocuous, though

Fig. 9.8 Shingle beach in south east England. Note the groins

unsightly, in cool weather it becomes soft and sticky in the sun. One way of dealing with this is to use a hydrocarbon-solvent-based, dispersant chemical which has been converted into a gell. Warren Spring Laboratory[7] have developed a suitable material and equipment which can increase the contact time sufficiently for it to deal with weathered oils.

Two knapsack packs are used, one with dispersant mixed with 20% of the gelling agent the other with sea water, and they both feed a single spray gun. This is sprayed on to the oily surface at the rate of seven square metres for one gallon of mixture. The recommended gelling agent is Triton + 100.* Suitable dispersants are listed in Appendix 2, only those shown for beach use should be used.

The treated surface should be washed down with sea water from a high pressure fire pump at or about high tide, so that the emulsified oil is washed into the sea. Once more it must be emphasised that the dispersant is probably more toxic than the oil itself, so that cleaning should only occur where rocks are part of a much used amenity beach or manmade structures subject to constant use (Fig. 9.9).

Mangroves

The effect of an oil spill on a mangrove area has already been discussed in Chapter 2. It will be appreciated that, not only can

Fig. 9.9 Man-made structure which is suitable for dispersal-gel treatment

the area be harmed by oil, but that any interference with area by the passage of amphibious vehicles and the like will also be highly damaging. In the event of an area of mangroves being contaminated it is suggested that nothing should be done (Fig. 9.10). If time allows and the equipment is available, deflection of approaching oil by a correctly deployed boom will clearly be advantageous. It may also be possible to remove oil which is still floating on the rivers and creeks which wind their way among the mangroves. No attempt should be made to remove oil from the stems, aerial roots and mud which form the lower strata of a mangrove area.

Arctic Conditions

Since the first edition of this book was published, 1976, the search for oil has extended both north and south into Polar regions. The successful development of the North Slope off Alaska, and the building of the trans-Alaskan pipe line has encouraged exploration in Northern Canada and the USSR, while exploration in the South of South America is extending still further around the Antarctic Continent. At first sight an oil spill, however produced, would in such surroundings give little cause for concern and it might be thought that remedial action would not be required; further consideration can change this view.

The single unique feature of these regions is the low average

Fig. 9.10 Mangroves

intensity of solar radiation and the wide seasonal variation which occurs. Much of the solar energy in the three summer months goes to melting the ice and snow, but does little to raise the general temperature. Consequently life is largely controlled by this. Vegetation develops flowers and seeds very quickly, and most animals and birds are highly migratory. The seas are very productive and consequently support special adapted larger mammals—seals and whales. Because of these conditions, the polar regions are both harsh and fragile. They impose great constraints on man's activities, but are easily damaged. There is a low species diversity, so the diminution of any one species could have a great effect on any of the others.

There is also an emotional issue. Many people regard these areas as the last great wilderness still unconquered by man and, seeing the mistakes which have been made by man throughout the ages in clearing the wilderness, be it the once productive deserts in North Africa, the dust bowl of the Middle West of North America or the destruction of tropical rain forests, feel that this last area should not be spoiled. Its riches could be utilised but the environment should be protected, unchanged.

With these varied concepts in mind much work, particularly in Canada, has gone into the problem of how to deal with oil spills on land, on ice and under the ice. Ice acts on oil as a diffusing mechanism. It allows it to spread on or below the layer and due to the low temperature, and consequent low evaporation rate, the oil persists in its toxic state for long periods.

The major problem in polar ice would be the effect of an oil well blow-out occurring at the end of a summer drilling programme and with oil escaping until a relief well could be drilled in the following summer. This has been discussed with particular reference to the Beaufort Sea, by Peter Wadhams[8] whose conclusions are worth quoting. 'It is clear that the extent of the oil spread from the blow-out, the persistence of the oil in a toxic state in the environment and, above all, the difficulty of clean-up are all considerably greater than for a blow-out in the open sea.'

The only feasible way of dealing with the oil is to attempt to burn it, when it first emerges through the surface ice and consequently still retains its lighter fraction. This will be difficult and dangerous to the lives of the operators. As the burning must be carried out shortly after the oil appears on the surface, aerial inspection and ignitors dropped from the same aircraft offer the only course of action, and in view of the large number of small random appearances of the oil, total success is most unlikely. When oil appears in the larger leads at the edge of fast ice or coastal strips, then booming and skimming might be possible. If the oil is present beneath fast ice, then the spread will be comparatively small, amounting to a few square kilometre, and the clean-up problem is easier.

If the spill is extensive and clean-up impossible, then there could be severe environmental damage, in particular to migratory sea birds which rest in the leads in spring and summer, to inshore fish spawning and nursery grounds, and to the coast line. The commercial value is small, but the real cause for concern is the damage to this, the last great wilderness on earth.

Protecting the shore

The problem of protecting the shore, that is, keeping the oil from getting to it, has been discussed in Chapter 5 on booms, from which it will be appreciated that this is almost impossible. However, in some circumstances it might be possible to use one of the surface-active chemicals, either to stop the oil from reaching the shore or, if it does, to reduce the clean-up problems. Tests carried out both in Heligoland and England showed that if sand or shingle was treated with the chemical before the oil arrived, then the oil did not sink or spread across the sand, and an incoming tide lifted virtually all the oil off again, in the form of compact thick slabs which drifted away. The sand beneath was comparatively clean. The tests suggested that the use of the material would greatly reduce the amount of removal or dispersal that was necessary and so simplify the urgent treatment of an amenity beach.

It should be appreciated that oil which is prevented from coating a beach by the use of such a surface active chemical will be carried away by the next higher tide and so is quite likely to be deposited, either back on to the same beach or on to another one. It will be decreasing in volume and probably increasing in both viscosity and density and accumulating sand and debris so that it may finish up as a solid deposit or tar balls.

Inland oil spills

This book is primarily concerned with oil spilt on to the sea and to a less extent to oil spilt on to rivers and inland water. Consideration should also be given to oil spills which occur inland. These are relatively common, although fortunately they are not large. They are produced by spills from the operation of filling the storage tanks at factories or at domestic heating plants or at inland oil-fired power stations. They also can be produced by the damage to oil storage tanks, either accidental or malicious, or by the spills resulting from damage to road or rail tankers carrying fuel oil or other hydrocarbons.

When oil is spilt on the ground, it will spread out and will almost inevitably sink through the surface if the ground is at all permeable. The depth to which it sinks, will depend both on the type and the structure of the soil and on the volume and, in particular, on the viscosity of the oil. Therefore, a low viscosity oil will rapidly sink into a dry sandy soil which is highly porous, but it will spread much more slowly through clay, and a waterlogged ground will resist penetration very much longer. However, in a case of this kind, a viscous oil will not sink in but will spread over the surface. In general, agricultural activities make soil porous for a depth of 10 to 20 cm, and oil will usually penetrate to this depth forming small underground pools where it meets damp or an impermeable layer. While this is occurring, the volatile components of the spilled oil will be evaporating, depending on the type of oil and the temperature of the ground and of the air. This evaporation may well produce a local fire hazard. Oil which penetrates the sub-soil will move downwards slowly until it reaches water or an impermeable layer. When it reaches ground water level, it will usually flow with the ground water and consequently a spill of oil can move a considerable distance below the ground. If the ground is fissured then the oil will move through the fissures as well. The maximum depth of penetration in porous soil can be estimated from the following formula:

$$D = K\frac{V}{A}$$

Where D = maximum depth of penetration (m); K = constant, depending on the retention capacity of soil and viscosity of oil, V = volume of infiltrating oil (m^3); A = area of infiltration at surface (m^2). Typical values for K of porous soils[9] are given in Table 9.1.

TABLE 9.1 Typical volumes of *K* of porous soils

Soil	*K* *Low viscosity petroleum products e.g. gasoline*	*K* *Kerosene Gas oil products with similar viscosities*	*K* *More viscous oils such as light fuel oil*
Stones, coarse gravel	400	200	100
Gravel, coarse sand	250	125	60
Coarse sand, medium sand	130	65	30
Medium sand, fine sand	80	40	20
Fine sand	50	25	10

The problem with movement of oil in the ground of this kind is not the presence of the oil but rather that it may migrate slowly in the direction of the ground water flow and so can eventually reappear in springs a long way from the spill, resulting in the contamination of the surface water. Alternatively, as in many places in the British Isles, water is extracted by means of deep bored wells and from porous strata under an impermeable layer, a major oil spill of a low viscosity oil can eventually produce some degree of contamination of potable water a long way from the neighbourhood of the spills. Even slight traces may be sufficient to give an unpleasant taste and so be objectionable in drinking water, while the presence of oil in larger quantities can effect the use of water for industrial and agricultural purposes.

In the event of an oil spill on land, immediate action should be taken to limit and stop the escape of oil, and to prevent the spilled oil from spreading. In particular, all possible action should be taken to prevent oil from reaching a water course or entering sewers or drains. Embankments of earth or sand around the spill to close access to road drains and sewers, should be put up as soon as possible, while inlets to drainage systems can be covered by using materials such as tarpaulins, blankets and plastic sheets. If possible, any visible oil should be picked up by means of a gulley emptier, or some other similar type of pump. Simple reservoirs, made of holes dug in the earth and lined with plastic sheets, can be prepared to act as temporary storage. If indeed the water does sink deeply into the ground, and endangers underground water supplies, then much more elaborate and sophisticated remedial measures must be taken which such as pumping and sinking a line of wells to capture the moving oil. These methods are outside the scope of this book.

Oil which enters streams or rivers can be dealt with by the various methods described in Chapters 5 and 6 and 7. It is in this type of situation that the use of absorbents becomes really practical. Absorbents, if finely divided can be thrown or shovelled on to small patches of floating oil in a river or pond, and can be removed either by an improvised dam lower down the stream, which will collect the oil-soaked material, or a by simple scoop system that can be used to remove the oil from the water surface. As, generally, the whole scale of the operation is small, it is consequently easier and it is possible to produce a very good clean-up. Booms, either production ones, or improvised booms made out of floating logs or bales of straw, as described in Chapter 5, can be used where the river flow is sufficiently low, or by putting them in at a sufficiently oblique angle.

It must be emphasised, that when oil spills occur inland, dispersant or similar chemicals should not be used. As already discussed, a dispersant only increases the surface area of the oil so that bacterial degradation can take place more rapidly. Although it is true that oil-degrading bacteria are to be found in soils and in inland waters, nevertheless, in almost all of these cases, the water is not completely saturated with oxygen. Bacterial degradation requires a great deal of oxygen so that if a quantity of oil is in a restricted volume of water, the abstraction of the oxygen, needed to degrade it, may be sufficient to reduce the oxygen level in the water below that which can support the natural wildlife. Many rivers are already rather short of oxygen due to the discharge of industrial and sewage effluents, so that any increased oxygen loss is to be avoided. Similarly in small areas, the chemicals, used as emulsifying agents, usually have some small degree of toxicity and their presence for long periods of time in lakes or ponds may well be sufficient to kill some kinds of wildlife.

If vegetation on the sides and banks of rivers, canals, and lakes,

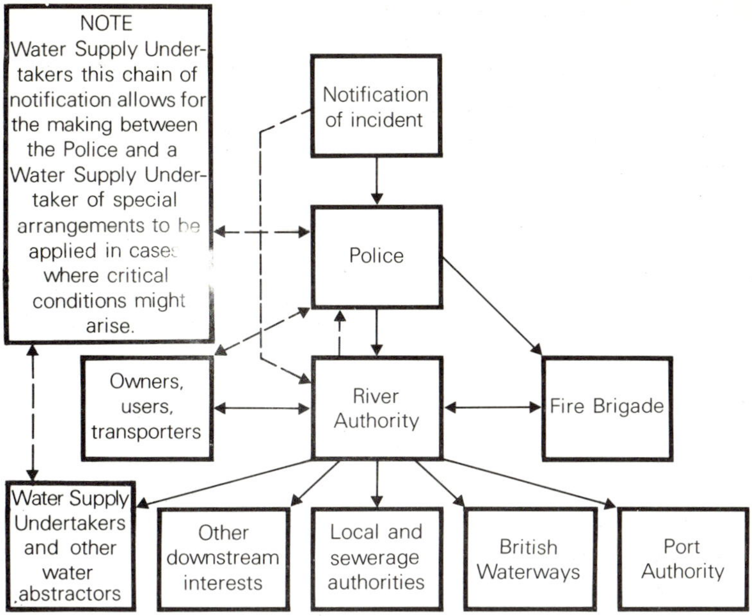

Fig. 9.11 Emergency treatment

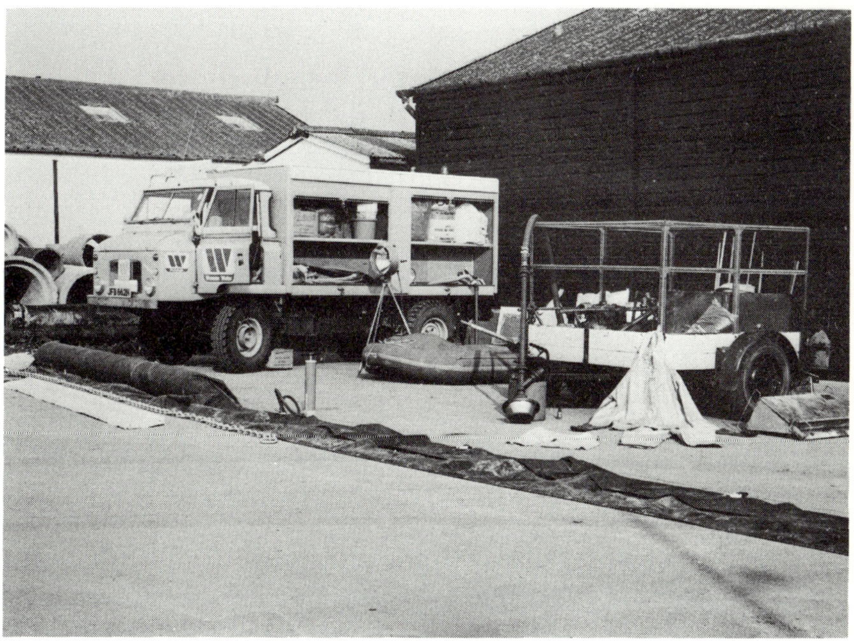

Fig. 9.12 Mobile emergency equipment

is oil-coated then the comments on the dangers of firing or cutting vegetation should be studied carefully. In general, burning should not be carried out, and cutting should only take place in autumn or late winter. At other times, cutting may damage the new growth more than the oil contamination.

Oil spills on land require a prompt and immediate action, so that all the various authorities concerned can be informed and take the necessary steps to minimise the possible effects of pollution. The Oil in Industries Working Group of the Institute of Petroleum have considered this matter in the United Kingdom situation.[10] Figure 9.11 is a diagram to indicate an emergency procedure which could be adopted to suit other situations.

An excellent example of good organisation has been set by the Wessex Water Authority who have made elaborate arrangements for emergency operation when oil spills and similar accidents occur. They have a night and day staff available on call and have equipped a van and trailer with what is, at least, a working minimum of equipment to suit a number of situations (Fig. 9.12). It includes such items as booms, skimmers, portable lighting, inflatable dinghy, ropes and anchors. While this system may not be able to cope with the biggest spills it is a step in the right direction, and could be expanded to meet a large number of situations.

Disposal of oil contaminated beach material

It is not easy to dispose of oil-saturated sand, shingle or absorbent materials which have been removed from the shore or banks of a water course. If the materials are piled into a heap it is almost inevitable that the oil will slowly percolate downwards and will eventually emerge, either to cause pollution of ground water or to re-enter the previously affected area. For example, when oil and oil-soaked debris were recently deposited in an abandoned gravel pit, it was found that the oil percolated through the strata and threatened a salmon river some little distance away. This problem is discussed in a Concawe Report.[12]

Before dumping any such material, the advice of the Local Water Authority or River Board should be obtained so that dumping can be undertaken with the least risk.

In Brittany, whose shores have suffered oil from the *Torrey Canyon, Amoco Cadiz, Tanio* etc., a good deal of work has been done on the treatment of this oily material and the potential uses of the treated material. The most successful has been treatment either with a hydrophobic quick lime or with a standard quick lime. Incineration was judged too costly and separation, or washing, too slow. In fact, lime was used on 165,000 tons of material between April and September 1978.

The basic treatment is to mix 7–10% of quick lime with the oily beach material, either by spreading layers on suitable flat ground and mixing mechanically using agricultural machinery (e.g. rotovate) or using a digger to mix the material in a plastic lined trench, or to reduce the lime dust problem, use mixing drums. The water in the oily material reacts with the quick lime and heats the mixture, thus aiding the mixing process. After treatment oil will not run out or be leached out by rain and the material is suitable for road filling, embankments etc.

Oily beach material, if not in too large a quantity, can be added to domestic refuse tips. In the United Kingdom the system of controlled tipping is used and the garbage is covered with a layer of soil every few feet. If the oily material is uniformly spread over the heaps it will be found to degrade fairly rapidly. The microbiological activity is already very high and, as the heap is porous, adequate oxygen is available. Oil is included in the general process of biological decomposition. As a consequence of this decomposition process heat is developed and the temperature can run up to as high as 60 or 70°C. In this process the oil will get much thinner and consequently will percolate directly through the heap, but the absorbent soil and the other material should prevent any run-off. It will be seen that this depends on the material being properly spread out and that an adequate amount of soil has been provided in the heap. If such dumps are not available then sites will have to be found where they will not interfere with any water catchment area, and in this case, too, the oily material should be spread in a flat layer so that the oil concentration is kept low in relation to the unit area.

This also allows adequate oxygen to aid microbiological degradation. If the material contains a very high percentage of oil then attempts should be made to get it accepted at the nearest oil refinery. Many refineries already have oil-water separators and are prepared to deal with other oily residues of various kinds.

References

1. The Beemer Beach Cleaner was developed for cleaning normal litter from sandy beaches, described in the Warren Spring Laboratory Report No. LR198. U.K. Agent Lacre Ltd., St. Albans, Hertfordshire, England. Overseas: Western International, P.O. Box 1123,

Pomona, California 91769, U.A.A. The M. F. Brighton (Australia) Beach Cleaner, described in Warren Spring Laboratory Report No. LR185. U.K. Agent: Massey-Ferguson Ltd., Coventry, England.

2. WIBAU Mobile Combustion Plant, Matthias & Co., Rothenbergen, W. Germany.
3. *A simple incinerator for the clean-up of oil contaminants on beaches: Design combustion and testing*, PACE Report No. 79–3 by Trecan Ltd., 4540 Dixie Road, Mississauga, Ontario, Canada. Published by Petroleum Association for Convention of the Canadian Environment (PACE), 1979.
4. TUNTZ, C. D. G., and MELOY, T. P., 'Froth flotation clean-up of oil contaminated beaches', EPA, API Prevention and Control of Oil Spills, Washington DC, 1971.
5. GREENWOOD, J. J. D., *et al.* 'Use of Waterproof limestone dust', *Public Cleansing* **LX** (4), 178 (April 1970).
6. FROST, L. C., ' "Torrey Canyon" Disaster. The persistent toxic effects of detergent on cliffed vegetation at the Lizard Peninsular, Cornwall', *Cornish Studies* **2**, 5 (1974).
7. NICHOLS, J. A., and LYNCH, B., *Dispersant gels for treating surfaces contaminated with residual oils*, Report LR 327(OP) Warren Spring Laboratory, Stevenage, Herts, UK, 1979.
8. WADHAMS, PETER., Oil and use in the Beaufort Sea—the physical efforts of a hypothetical blow out, *Petroleum and the Marine Environment*, Graham & Trotman, London, 1981, pp. 299–318.
9. NIGHTINGALE, J., and NICHOLS, J. A., *Beach Protection with Shell Oil Herder*, Report LR 184 (OP), Warren Spring Laboratory, Stevenage, Herts, UK, 1973.
10. *Oil Spill Control Manual*, Shell International Marine Ltd., June 1974.
11. *Inland Oil Spills Emergency Procedures and Action*, Oil & Industries Working Group, Institute of Petroleum, London, 1972.
12. *Disposal techniques for spilt oil*, Concawe Report No. 9/80, Concawe, van Hogenhoucklaan 60, 2596 T. E. den Haag, The Netherlands, 1980.
13. BELLIER, P., *Treatment of oily wastes*, CEDRE Report, France, January 1979.

Part Four

Organisational Aspects

10

Forward planning

Planning and organisation

The general problem

As we have already seen, over the last 50 years, the number and size of oil spills has considerably increased. At first, spills were small, and, more important, the general public was not particularly concerned about the environment. People grumbled about oil on the beach but presumed it was, in some way, the price of progress, or of having an oil terminal or refinery, which was, of course, a valuable employer of local labour in the area, so that, in general, no effort was made either by the spiller of the oil or by the public authorities to clean up beaches. Public opinion has slowly changed, protection of the environment is no longer confined to a few naturalists, and many organisations, particularly those concerned with bird life, have protested vigorously both against spills where birds have been killed and against the threat of spills from proposed new installations. At the same time, the amount of oil carried by sea has increased and with it a corresponding number of spills, many of them very large. All of this resulted in the IMCO conventions which have been discussed earlier. National or international legislation will not automatically reduce the amount of oil spilt by accident, either by the tanker at sea or while it is loading, discharging or bunkering. Spills of this latter kind, as they occur at oil terminals or in port, are usually closer to centres of population and therefore experience of this pollution has led to an increasing demand by the public for the pollution to be removed.

The methods to use and the type of equipment have all been

discussed earlier. This chapter deals with the organisation which is needed if this equipment is to be used properly and effectively.

Western Europe Practice

In many countries, particularly in Western Europe, oil spill clean-up is undertaken directly by the State. For example in the United Kingdom, the task of dealing with an oil spill at sea is carried out by the Marine Branch of the Department of Trade, which deals with all maritime matters.

Following the *Amoco Cadiz* and *Eleni V* incidents, the Department set up a Marine Pollution Control Unit (MPCU) within the Marine Division.[1] This Group is responsible for selecting the best equipment to use and seeing that it is provided. They maintain an information/control room in London and, in the event of an incident, they assume control of all operations at sea. For administrative purposes the Department has divided the coast line into eight areas. In each there is a Principal Officer who, with his staff, is based in the largest port in the area. Attached to each area is a Marine Pollution Control Officer who is responsible to the MPCU in London. The Coast Guard, also part of the Marine Branch of the Department of Trade, is similarly organised.

Oil spills are reported from many sources, from observers on the coast, coast guards, police and private citizens and from aircraft flying over the sea (both military and civil aircraft are requested to inform their controls if they see any large oil slicks). This information is passed to the Coast Guards, who coordinate it and then report to the MPCU, who decide what action to take, sending one of their staff, usually the local Marine Pollution Control Officer, to the scene to obtain more precise information. The Marine Rescue Coordination Centres, part of the Coast Guard organisation, have adequate communication facilities and are used, if required, as a local control room for the incident.

The method adopted, at present, in the United Kingdom is to disperse the oil and equipment for spraying dispersant is available all round the coast, together with stocks of low-toxicity dispersant. Prior arrangements have been made by the Marine Pollution Control Officers in each area, so that in the event of a spill, ocean-going tugs from local ports are fitted with the equipment and go to deal with the oil. All the tug crews have been trained, and regular large scale exercises are carried out. The loading of dispersants and the fitting of the equipment can be quickly carried out and the tugs can soon reach the oil. The use of aircraft to spray the newer, concentrated dispersants is being examined and specially fitted tanker aircraft may soon be available. Sets of the Spring Sweep equipment (see p. 148) are also available. The Department of Trade has established stock piles of specialised transfer pumps and associated equipment to assist in the off-loading of the cargo of any damaged tanker.

However well and expeditiously oil is dealt with at sea, some of it is likely to reach the shore. In the British Isles, this oil is cleaned from beaches where it would interfere with amenities, by the local authority, that is, the local government organisation which deals with street cleaning, garbage collection and so on. These authorities generally use their own labour and have stockpiled material and equipment, which have been approved by the responsible Department which is the Department of the Environment. In the event of a major spill, this Department will set up a Coastal Pollution Co-ordinating Centre to act as a focal point for information transfer, coordination of clean-up strategy, assessment of priorities and allocation of resources. They also hold a stock pile of, mainly, shore cleaning equipment intended to supplement municipal resources.

In the event of a large spill from a known tanker, the cost of clean-up is settled by a government grant, the government itself being recompensed by the tanker owner, or by the owner of the refinery which has spilled oil, either as a legal liability or under the Tovalop agreement. Procedures of this kind are adopted by a number of countries. The important factor is that the state takes over the direct responsibility of cleaning up and does not employ agents. When they are satisfied that the area is clean enough, the operation is stopped and the local authorities are confident that the direct cost of their clean-up will be met by the spiller of the oil.

United States practice

Legislation in the United States, and in a number of other countries, is rather different. In these cases, the spiller of the oil is responsible for the clean-up. The United States Government designates an officer as the on-scene commander. For spills at sea, he is a member of the United States Coastguard, and for spills on inland waters he is a member of the staff of the Environmental Protection Agency. This officer's job is not to carry out the work but to oversee it and to agree with the body doing the clean-up that a satisfactory job has been done. In the event of the actual spiller of the oil being, for some reasons, unable to clean up the mess, or fails to have the work carried out, then the department undertakes the work either directly, or by employing a contractor, and finally sends the bill to the spiller, who has a legal obligation to pay.

The organisation which is actually handling the oil and, consequently, is the potential spiller, be it a refinery, a terminal or a tanker owner, is not necessarily directly equipped to deal with spills. In some areas, even if he were able to do so, there would be duplication and waste of effort where many terminals and oil-using installations are situated close together. Consequently, in many areas, oil-spill cooperatives have grown up (there are well over a hundred of these organisations in the United States alone). In general terms, these cooperatives follow a very similar pattern. Firms in the area who receive, or send out, oil by tanker or barge or in some cases who store it on the water front, agree to pay an annual levy, usually based on their oil movements, to a central fund. This money is used to finance the purchase of oil-spill equipment and to pay the cost of labour which is employed all the year round and is always ready to deal with an oil spill. All this is usually organised in one of two ways, either a clean-up company is formed whose directors are members of the cooperating firms and who employs a manager, and industrial staff, purchases equipment and sets up depots in the usual way, or, alternatively, a contractor is paid for the service. Any new specialised equipment required is purchased by the cooperative group and allocated for the use of the contractor. In the event of a spill from one of the member firms, the clean-up team can be quickly on the scene and is helped, if necessary, by the staff of the member who has actually spilled the oil. The equipment and indeed the serivces of the men are usually available for hire to non-members in the area. There are, of course, many individual variations of this basic idea.

The methods outlined above can operate satisfactorily within almost any country. Problems, however, can arise where installations are close to international boundaries and where an oil-spill can occur across such a boundary. Government clean-up organisations, however good, cannot operate, without permission, within the territory of an adjacent sovereign state, that is, of course, without a good deal of prior negotiation and an actual request for help at that particular time. This point was realised in 1969 by the countries around the North Sea who met and signed the Bonn Agreement. This specifies that details of the movements of a large spill from one country's zone towards another country's zone is given as soon as possible. Each country will deal with a spill in its own zone, but neighbouring states are ready and willing to help if the request is made. In fact, equipment and men from the Netherlands have assisted in at least two spills in United Kingdom waters.

The organisations already discussed, in general, apply to relatively small or homogenous areas. For example an area such as San Francisco Bay, or the Southern Coast of England, can be covered by a single well-placed unit which can move rapidly to a spill anywhere in that area. When, however, distances have to be measured in hundreds, or even thousands of miles, then in some cases it would mean a very long wait before equipment would arrive to deal with an incident and, as is well known, speed is the essence of good and economic clean-up. In cases of this kind, there would seem to be two alternatives. The first would be to provide first line equipment at as many ports as possible and specify that, if a spill occurs, then the nearest equipment goes out and starts to deal with the oil, and more sophisticated or additional equipment is moved in as soon as possible from adjacent areas. The second alternative, as air transport becomes more and more available, and many otherwise almost inaccessible areas have good air fields, is that, by choice of suitable aircraft and equipment, a speedy service can be provided from a central point, even over great distances. This is demonstrated in Fig. 10.1 which shows relative times for a 10 mph ship and a 300 mph aircraft to travel long distances, assuming that they both take six hours to load up. Unfortunately, this simple concept is not based on reality. Suitable aircraft are not standing-by waiting for a charter; they may well be elsewhere or have a prior contract to fulfil. The provision of such aircraft and crews standing-by waiting for an oil spill is too costly to be envisaged by any country. When aircraft have arrived or are available, servicing, crew rest time, and then loading can take as long as two or three days. When the plane and its cargo arrives at the airfield nearest to the incident, the equipment has to be unloaded and got to the actual position of the vessel. This may well include transfer firstly by road to a nearby port and then finding a suitable vessel and putting the equipment on board. Table 10.1 indicates the sort of times that this may take. It will be seen that even in quite favourable circumstances, this could easily take 20 h and under less favourable circumstances, it might take as long as 84 h. Many of these problems can be avoided if aircraft are designated for this task and are only used for oil-spill clean-up. The cost of maintaining aircraft and crews is so high and in any given area oil spills are so relatively infrequent, that this is not a practical solution.

Even if there is not a cooperative, it is customary in oil terminals

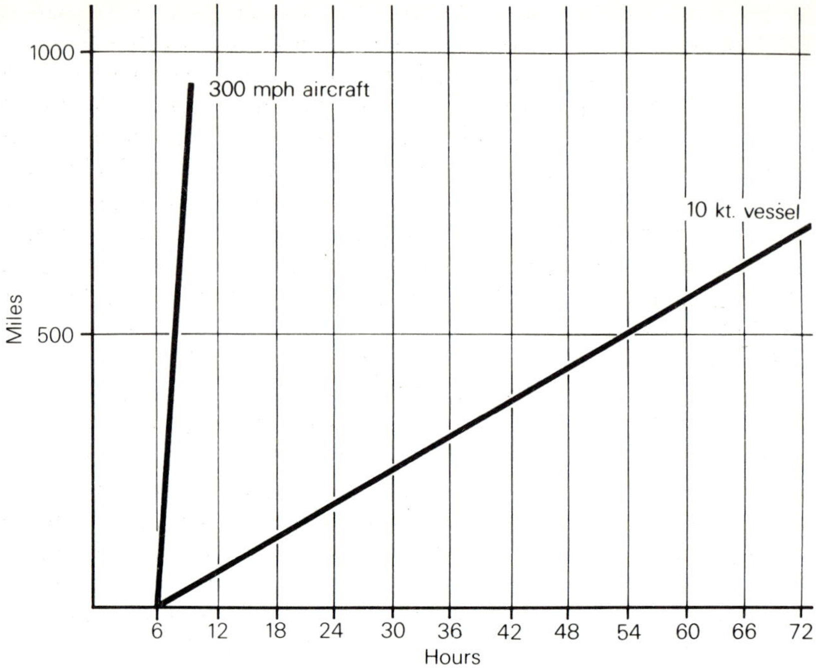

Fig. 10.1 Relative times for ships and aircraft to travel long distances

TABLE 10.1
Relative times for ships and aircraft to travel long distances

		TOTAL TIME	
		max.	*min.*
Move equipment to air field	6 hrs	6	6
Wait for plane to arrive	0 hrs up to 48 hrs	54	6
Service and load aircraft	1 hr up to 12 hrs	66	7
Flying time including weather, diversions, etc.	say 4 hrs	70	11
Unload	2 hrs	72	13
Truck to harbour	2 hrs	74	15
Load onto chartered vessel	1 hr to 4 hrs	78	16
Sail to scene, say 50 mls., at 8 kt.	6 hrs	84	22

to have an arrangement whereby the operator can deal with the normal 'accidental small spill' which inevitably occurs from time to time. The minimum size of spill which is treated will depend both on local geography and environmental situations and on the attitude taken by the local state in enforcing their national law. The equipment and arrangements will be tailored to deal with the local situation as these arrangements are only intended to deal with a relatively small spill, when the tanker is actually at the jetty or an offshore buoy, or very close to it. A large spill, or one outside the immediate local radius of operation, is not dealt with by these arrangements.

In some places there are no large maritime or industrial centres yet there are a number of oil terminals, for example, in the Caribbean Sea. Oil companies are accustomed to co-operate on various non-commercial aspects of their business, so it is suggested that they could agree to co-operate in a regional clean-up scheme. In brief, the method proposed is to divide the sea area concerned among the larger terminals of the oil companies in that area. Each of these installations is responsible for any spill in its area, no matter who is the owner of the spilled oil or which tanker or installation is concerned. To do this, their small scale equipment will have to be augmented with equipment suitable to deal with larger spills anywhere in their area, and the refinery or terminal personnel will have to be made available, and indeed trained, to operate this new equipment. These different installations are expected to cooperate with each other across the whole of the area concerned, or within parts of it if it is too large. Each area or zone should have a coordinator, who in the event of a spill, will be able to call on adjacent installations for further help. It will be apparent that this may not be particularly easy to organise as a company is expected to send its equipment away from its own installation to help to clean-up a spill in which they have no direct interest. Nevertheless, this can be and has been done in some areas of the world, and represents an efficient and economic solution to the problem.

Contingency plans for specific countries or areas

Every area in the neighbourhood of an oil installation and also where tanker traffic regularly passes along the shore-line should have a contingency plan. It will be appreciated that this means almost every sea coast in the world. It is emphasised that these contingency plans must be local. They must be drawn up by local people who have an accurate and intimate knowledge of all details of the areas and of every aspect of the problem. Nevertheless, these contingency plans have a considerable similarity and there are a number of basic facts which must be considered and included. First, what is the risk of an oil spill? This must be determined by examination of the amount of oil which is moved into or out of the area and the amount of oil which is carried close to the shore concerned, i.e. the number and sizes of the tankers. The past record of the area of oil spills will be most helpful in this respect. The next point to consider is the maximum probable size of oil spill. This is not the absolute maximum size which would, in the limit, be the size of the largest tanker that could possibly traverse the area. It is rather, what is the best estimate that can be made of the largest spill that is likely to occur. If the oil is from a well, then what is the maximum rate of discharge, presuming that the control devices were inoperable, and how long would it be before the flow could be checked or stopped altogether? In the case of tankers, what are the sizes of tankers which most commonly pass through the area? What is the likelihood of collision between tankers or between a tanker and another vessel? What are the weather conditions? If a tanker grounds, will it be on rock or on sand? Is the vessel likely to break up or be dragged off unharmed? Careful consideration of these factors, together with a scrutiny of past records, both in the particular place and in similar situations throughout the world, will enable the estimate to be made of the maximum probable size of spill. It is essential to try to do this exercise and to produce a figure, because if an organisation is to be set up, it must have a target. It must be able to deal with so many thousand tons of oil in so many days. If the spill is smaller than anticipated, then obviously the clean-up will be quicker and more effective. If the spill is larger, then either the clean-up will take longer or, alternatively, additional equipment will have to be obtained from other places, flown in or brought in by sea.

The next point to be considered is why must the oil be cleaned up, i.e. what is at risk and what would be damaged if nothing was done? It will be appreciated from Chapter 2 and 3 that, in many circumstances, there is much merit in doing nothing to the oil, but leaving it to nature to remove. Nevertheless, there are also many conditions in which some action must be taken. If the oil threatens amenities, when they are actually being used, or just about to be used, then, obviously, it must be removed. Similarly, if oil is threatening a sea bird area when the birds are actually present or just before they are due to arrive, then again, the oil must be cleaned up. But it is suggested that, if oil arrives in winter, on a fairly remote coastal area largely used by holiday makers in summer, then, providing it will have been removed by natural causes before the start of the holiday season, there is little reason why any clean-up should be carried out.

In dealing with the shore clean-up problem in particular, it is essential that local advice is taken from all of the scientists who are concerned with wildlife in that area. In the United Kingdom this will consist of the Nature Conservancy, the National Environmental Research Council, the Ministry of Agriculture, Fisheries and Food, the local Sea Fisheries Committee, and in some areas the Agricultural Committee or the local Farmers Union. From the advice received from these various bodies, a map must be prepared of the coast line on which will be marked various areas of particular interest be it birds nesting there or shell fish beds (Fig. 10.2 gives part of the map of the coast of Kent as an example). Having marked these areas, a clean-up strategy can be developed. Some areas may have to be cleaned immediately, whatever the time of the year. Other areas, perhaps should never be cleaned at all because the potential danger due to cleaning, or even by men walking or driving vehicles on the foreshore is too great to be permitted. Other areas may have to be cleaned during the summer months having been omitted during bird breeding seasons. In some areas dispersant, even low toxicity varieties, should not be used, in other areas it is essential to get the beach clean no matter by what method.

In the USA a rather similar system, more suitable for extensive areas, is being developed.[2] It consists of ranking coastal environment on a scale of 1 to 10 with respect to potential oil damage. Generally speaking areas of high wave energy or tidal current are low, whereas sheltered areas, where oil can accumulate, are high. That is exposed cliff and rank areas would be I, while sheltered areas would 8 or 9, and saltmarsh 10. To these numbers are added details of coastal fish and wild life,[3] together with amenity and industrial areas. The basic numbers are an indication of the need to remove oil or, better, to protect the area from the oil—the higher the number the more important.

All these discussions, and the drawing up of this plan, must be done before the oil arrives. It cannot be left until the threat of pollution is imminent because innumerable mistakes will then be made, the cost of clean-up will be increased, and the potential danger to wildlife is almost certain to be greater.

When all these considerations have been taken into account, then it will be possible to determine how much equipment is

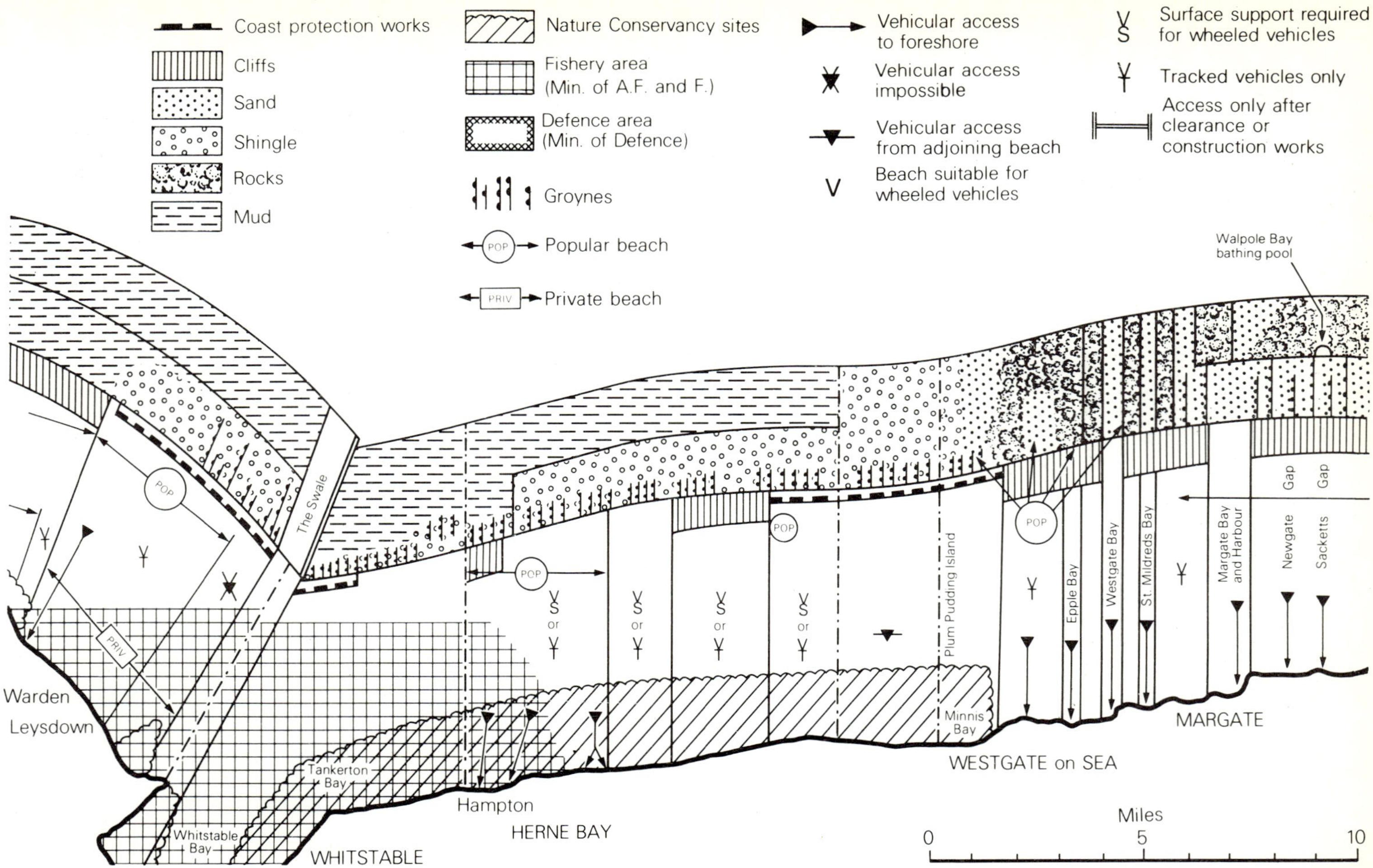

Fig. 10.2 Part of a contingency plan for the county of Kent, England, showing beach type, use and access points

required to fulfill the requirements. This can be divided into: equipment to keep the oil away from the coast, for example booms or chemical herder; equipment to disperse the oil at sea before it reaches the coast; equipment to contain and pick up the oil, and then, when the oil comes on to the beach, earth-moving equipment to remove dirty beach material, oil pick-up equipment, hand held and other types of dispersant equipment, pumps, sea water hoses, etc. Stocks of expendable material must be purchased and stored. Above all, an organisation must be set up to deal with the spill, men must be appointed, and men must be trained to use the equipment.

It is essential that, if oil is spilt, all of the contingency arrangements move into action smoothly and rapidly. Time is the essence of the whole operation.

Organisation

In setting up an oil pollution contingency plan to cover the shore lines of a whole state it is essential that all the various units can be centrally controlled and coordinated so that, in the event of a major incident, all of the resources can be utilised to best advantage. The arrangements suggested at the IMCO Symposium at Acapulco in 1976[4] are convenient and flexible. The paper points out that, from the point of view of clean-up operations, oil spills can be divided into three groups:

1. *Chronic Pollution* which, in this context, can be described as the tarry lumps and occasional patches of viscous semi-liquid oil found on almost all coasts and particularly on windward ocean shores.
2. *Medium Spills* which are mostly produced by ships during handling of their bunkers and/or cargo and almost always occur during transfer operations. These spills by their very nature are usually small and constitute 97% of all spills from tankers.
3. *Large Spills* can result from a major incident involving a tanker. Although this type of spill is infrequent, only 3% of all spills, they are the ones that can cause serious pollution. In two-thirds of these incidents more than 50 barrels of oil entered the sea.

Combatting pollution requires planning at these three levels to be complementary and, as the size of the spillage increases, the need for co-operation between levels at national, and sometimes international level, becomes essential. Oil knows no boundaries, and the plan must satisfy the basis requirements for the rapid, controlled and coordinated reaction from properly trained men using the right equipment. In formulating the plan, it is essential to ensure that the clean-up of each and every spill, whatever its size, is efficient, i.e. cleaned up with the least damage to the environment and at the least cost.

The contingency plan will have to be written or approved by the responsible authority for the country or area concerned. It may well be divided into several parts, but the following fundamentals will have to be fulfilled by each and every part.

1. The plan must be clear, simple and include the details of the chain of responsibility.
2. At all three levels there must be immediate and effective response. The men, equipment and material required must be available at the scene of the pollution.
3. Logistic support and reinforcement must be readily available.
4. The man at the scene in charge of operations must be able to devote all his energy to the clean-up task.
   The coordination, provision, supply and reinforcement of men and materials, and all other related responsibilities, must be carried out by others.

In general terms, the 'Chronic Spill' will be cleared up by people working on the shore, the 'Medium Spill' in harbours and adjacent waters, whilst a 'Large Spill' may well have to be dealt with at sea, inshore waters and on the beaches. This is shown diagramatically in Fig. 10.3.

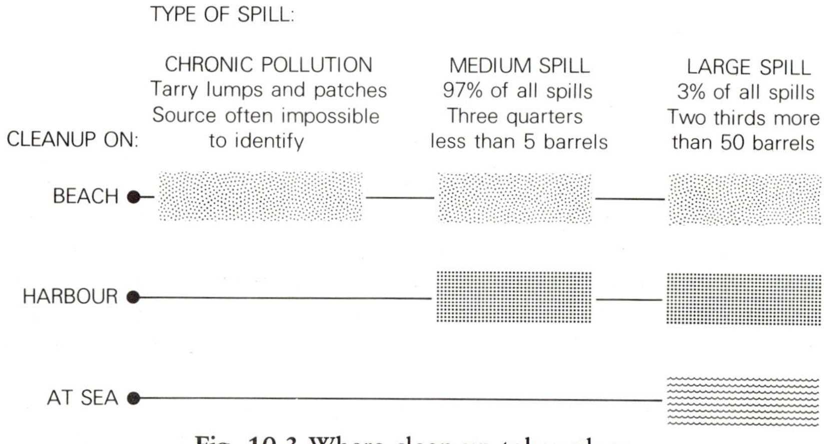

Fig. 10.3 Where clean-up takes place

The control and organisation required to clean up chronic pollution is outlined in Fig. 10.4. This is frequently a routine operation, mainly carried out on amenity and pleasure beaches. Warning of the oil on the beach can come from many local sources, and some formal arrangement is necessary for channelling such information to the right quarter. Alternatively, warnings, directives or requests for help will be received from a higher authority in the chain, in which case the local organisation becomes responsible to that authority for all clean-up operations carried out. When dealing with chronic pollution, the administrative supervisor responsible for this unit will decide when clean-up is necessary and when it has been adequately carried out.

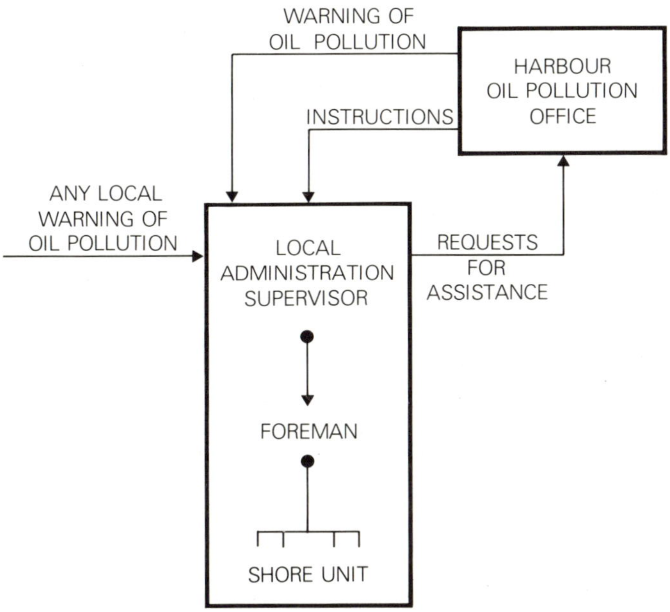

Fig. 10.4 Shore unit organization

Medium spills, occurring mainly in harbours, will be dealt with by a harbour clean-up including both boat and shore parties. Figure 10.5 shows the outline organisation required to clean up this size of spill and its place in the chain of responsibility. Warning of oil pollution can be seen to come from either local reports or higher authority as described above. If required, this organisation will control the shore clean-up gang through their local supervisor.

Figure 10.6 suggests a minimum organisation which would be required in a port during an incident. The increased threat from

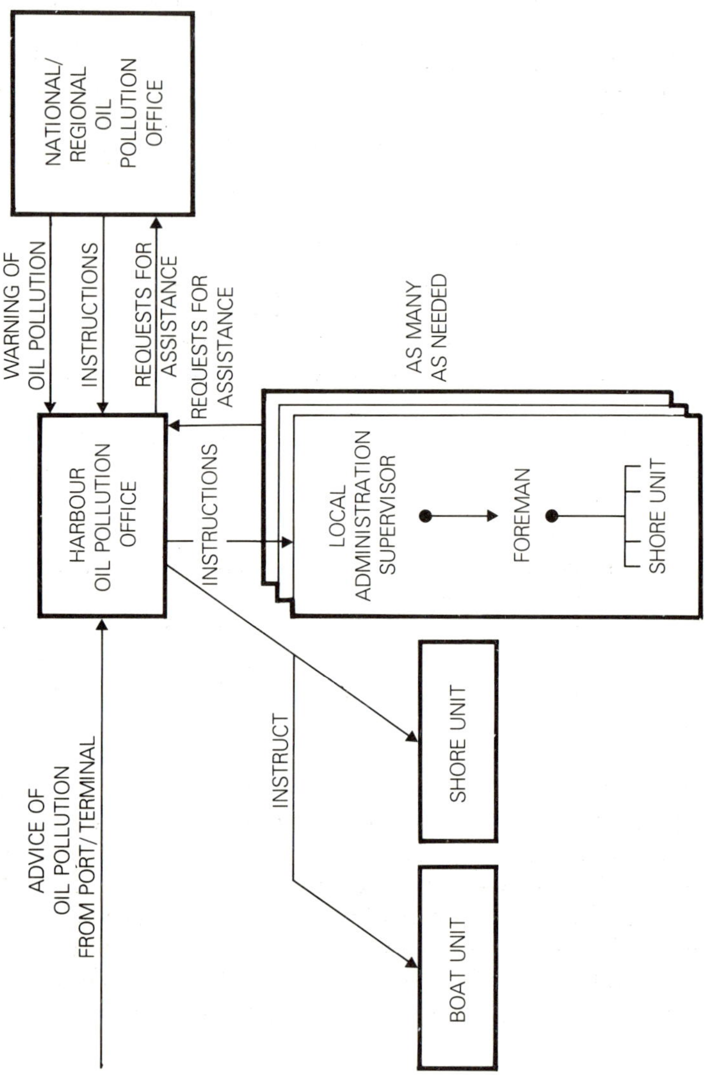

Fig. 10.5 Organization of harbour clean-up

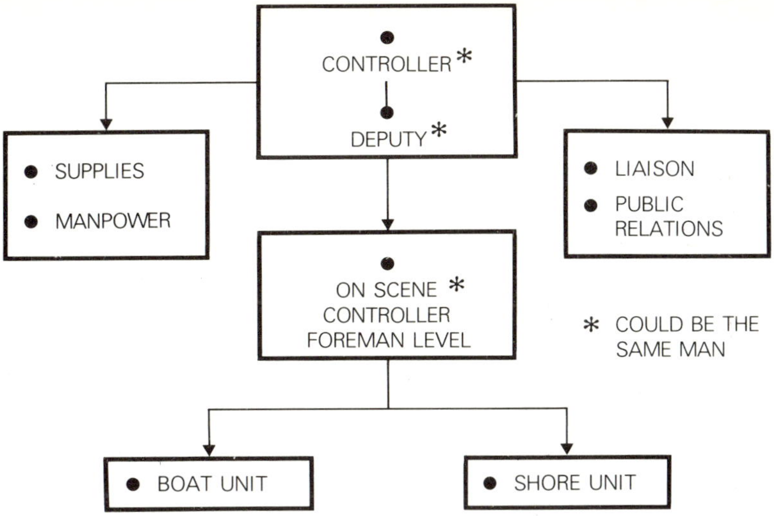

Fig. 10.6 Port/terminal oil pollution office

this size of spill requires more comprehensive planning and clean-up capability. The organisation can either be the responsibility of the local port authority or oil industry, as most applicable, and is simply a much scaled down version of the organisation for large spills described below, e.g. a trained oil pollution coordinator will be needed, possibly on a full time basis, with an oil pollution control office from which to coordinate clean-up activities. He will require assistants both ashore and afloat. These may well be people who normally work at similar tasks in the port or oil terminal.

Combatting pollution from large spills will always require coordination and control at the highest level. No country can afford to set up an autonomous organisation solely for this purpose. The task could be equally well carried out by a Central Agency which coordinates the work of other Departments and Services. A National Oil Pollution Office could be the central agency for coordinating and controlling all oil spill activity throughout the state and would delegate authority and responsibility to regions should this be required. This is shown diagramatically in Fig. 10.7.

Central Control

In any area, large or small, one man must be responsible for the whole of the spill operation and must be available day or night. Consequently, a series of men will be needed and the organisation

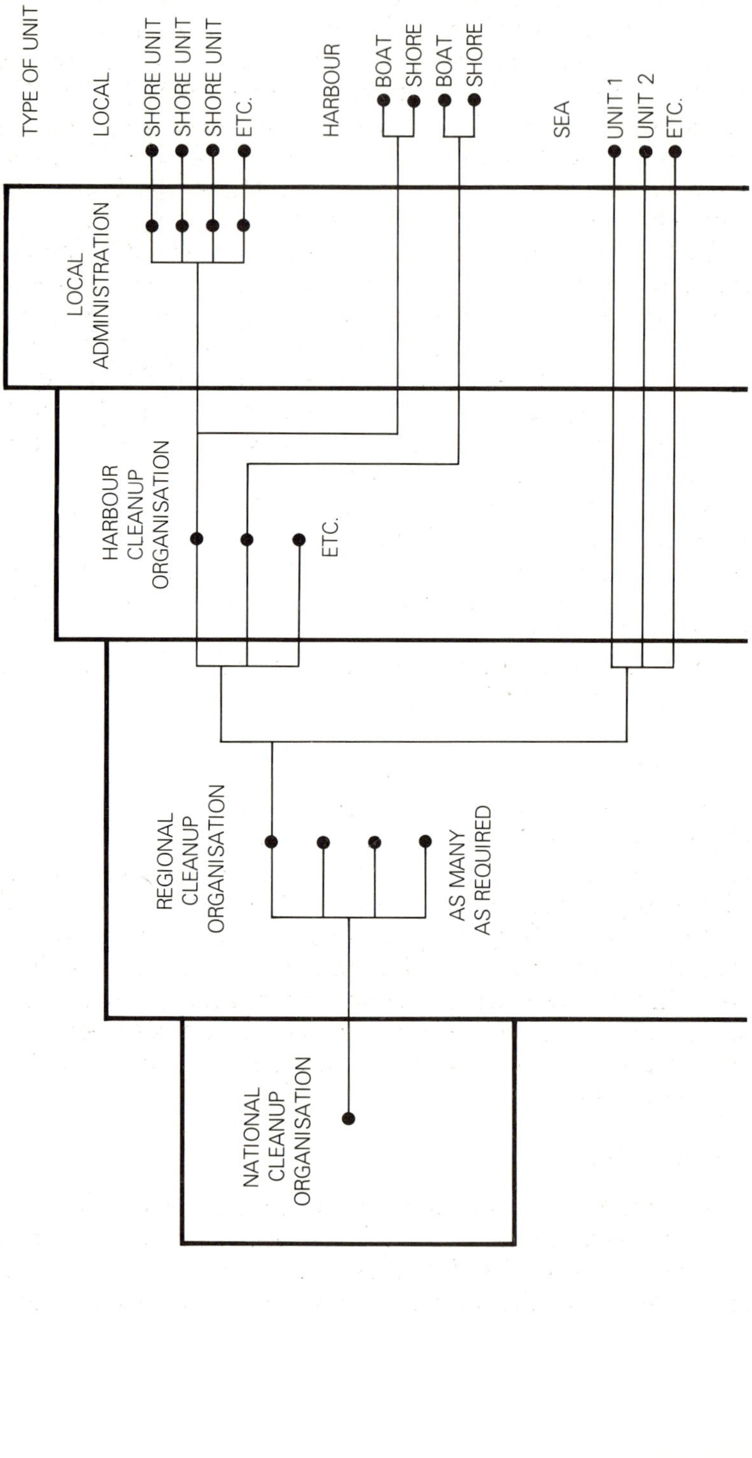

Fig. 10.7 Outline of a national organization

must be such that if one is away from duty for any reason then the next man is automatically on duty and automatically found by the oil alarm system. This man, whom we will call the controller, must have complete authority to deal with any matter which might hamper the clean-up operation. A single man can obviously despatch a well trained team to deal with an incident, and can get over the first few hours, but plans must be made for the operation to continue for much longer. It is, therefore, proposed that an organisation such as that shown in Fig. 10.8 should be set up. There are five functions to be carried out, either by one man or by five separate individuals, or even groups of individuals depending on the scale of the operation, who will be able to deal with the functions of on-scene controller, public relations, supplies, manpower and liasion with oil companies. It is anticipated that most of these posts can be filled by men who are already carrying out a very similar task in the day-to-day operating of the terminal or of the local authority department who are responsible for the total organisation.

The on-scene controller will be responsible for sending orders to the on-sea clean-up units and to the harbour, or regional, control offices as the case may be. The flow of information from that office and of instructions from it, during a major incident, are shown in Fig. 10.9.

The functions of the public relations, supplies, manpower and liasion officers are given below.

Public relations

It is essential that rumour is not allowed to spread. Therefore immediate and authoritative statements must be made. The man in charge of public relations will know the channels by which to deal with press, television and radio, and he will put out as rapidly as possible, and as frequently as possible, bulletins describing what has occurred and the action that has been taken. It is appreciated that at the start of the spill it is almost impossible to say how it occurred, and how much oil has actually been spilt. Nevertheless, as soon as possible, accurate information should be given to the public.

The supplies office

If the operation goes on for any length of time, then undoubtedly some of the special materials employed will be used up or stocks will become too low. The Supplies Officer, who may well be a purchasing officer in the oil company terminal or local authority,

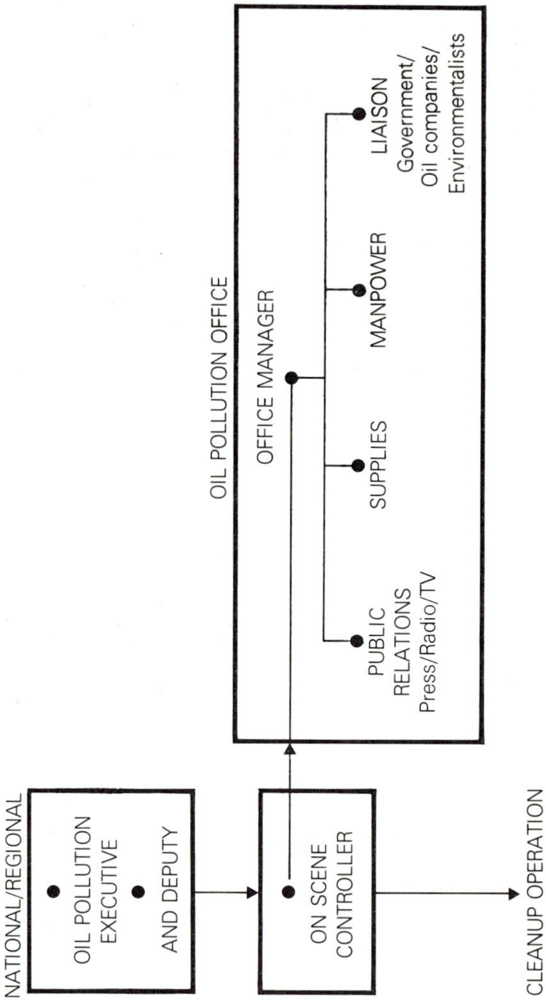

Fig. 10.8 Top management structure required

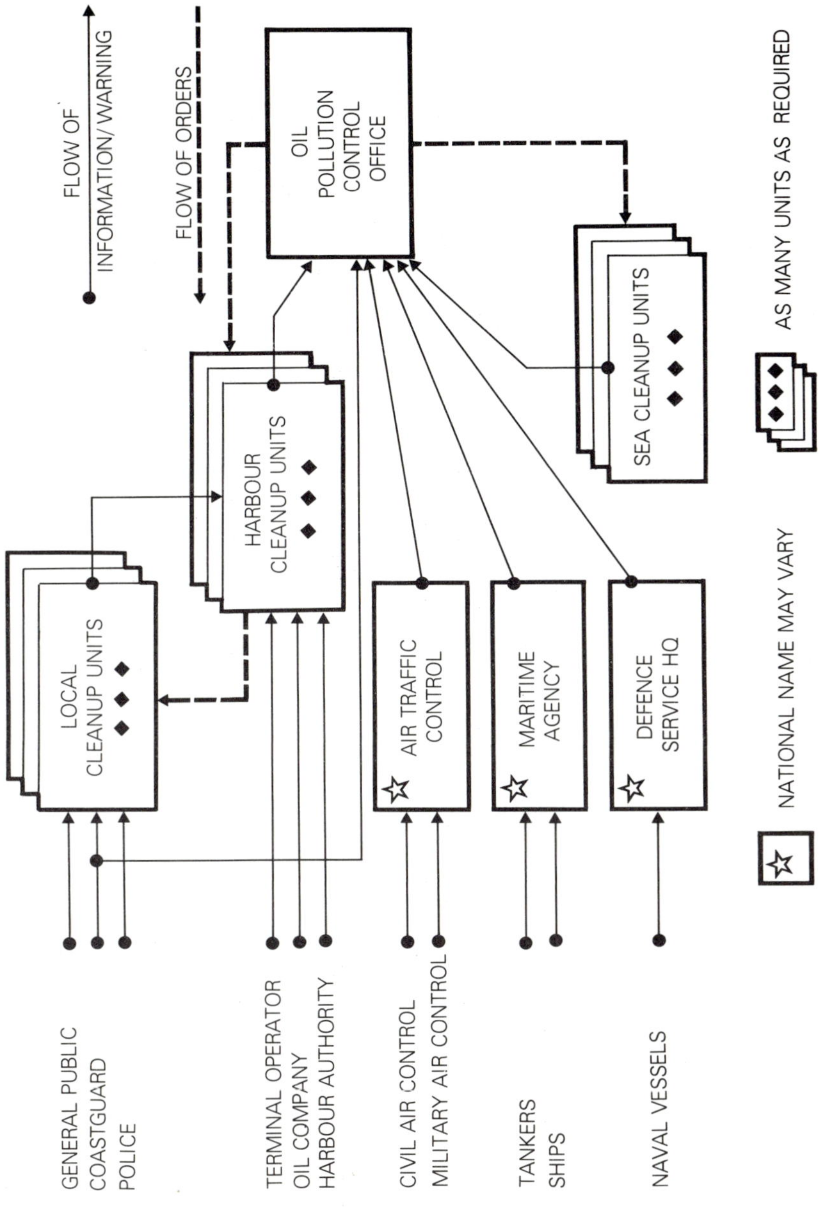

Fig. 10.9 Communication network

will have, in his office, all details of where to obtain the special items of equipment, and in particular, where and how they can be obtained, over week-ends and public holidays. He will be in a position to check the stock position continually, and find out the rate at which supplies are being used so that no material runs completely out of stock.

Manpower office

If the emergency lasts for any length of time, then some of the men will need to be replaced. They cannot be expected to work for more than perhaps twelve hours straight off, or for more than six days continuously. The manpower office will be aware of this and will do their best to make relief available.

Organisation of the operation

A description has been given of the type of organisation which should be set up in any area to deal with the threat of an oil spill. There are a number of features of the organisation of the operation itself which have been learnt from past experience, which will now be described.

Warning of the spill

Information about an oil spill can arrive in many forms—from the vessel or terminal spilling the oil, or from the terminal itself, from observations made by the Port Authority, by coastguards, or by members of the general public, and reports of oil spills seen at sea can be obtained from aircraft, either military or civil, landing in the country concerned. Arrangements must be made for all the different warnings to be funnelled through the same channel so that the controllers office will hear of them as soon as possible.

Operations Room

The controller must have an operations room. This can be at a static point, in a mobile caravan or even on a suitable vessel. It must contain adequate communication equipment so that all of the units concerned with this clean-up, whether on sea or land, can be reached with the minimum of time. Details of weather, tide and sea state must also be available together with all the maps of the sea and the land, and the details of the local clean-up methods discussed earlier. The other officers, or offices of the organisation, described above must also be in very close contact with the operation centre. When the controller has been told that oil has been spilt, he must immediately decide what to do in view of the state of the tide, the wind direction and force, where the oil has been spilt, the estimated quantity and where is it likely to go. From this information he can determine which of the equipment at his disposal should be dispatched to deal with it, what shore authority should be informed and whether neighbouring countries should also be informed of the imminent threat. As soon as this decision has been made, the various teams can be alerted, and the equipment sent to go about its task. (Figure 10.9 shows the flow of information and instructions into and out of the Controller's Office.) The need for adequate communication cannot be stressed too highly. Too frequently in the past it has been found impossible, except by an indirect route, for the controller to speak to ships or even to aircraft. This must be avoided as far as is possible. There is also a requirement for aerial surveyance. If the incident is a large one, then, at the very least, morning and evening area survey by competent well trained pilots must be carried out. This is an essential part of the whole process of dealing with the incident.

Training

Even on a small scale, the organisation to deal with an oil spill is not simple and depends on the cooperation and inter-relation of a number of individuals. Training, in all its aspects and at all levels, is therefore essential. Men working particular pieces of equipment must know how to operate them, which means regular exercises, with every piece of equipment and every man, must be carried out. The Controller and his staff must also be thoroughly experienced, and it is suggested that after the organisation has been set up, a number of paper exercises are carried out to see that everything works smoothly and that everybody knows what he is doing. If possible, at least one annual exercise on a practical scale should be carried out in which an imaginary spill takes place in a given sea area and a certain amount of the equipment is deployed to deal with it. This will be particularly valuable in assessing the capabilities of all the communication equipment referred to above.

References

1. *Improved arrangements to combat pollution at Sea*, Department of Trade, London, 1979.
2. MILES, O. H., GUNDLACH, E. R., and GETTER, C. D., 'Sensitivity ranking of energy, port shorelines', Proc Speciality Conference on Ports, Norfolk VA, ASCE/May 19–20 1980.
3. GETTER, CHARLES D. *et al.*, 'Mapping the distribution of and valuable oil-sensitive coastal fish and wild life, Oil Spill Conference, American Petroleum Institute, 1981, p. 355.
4. WARDLEY-SMITH, J. and GARNETT, M. J., 'Contingency Planning', Symposium on prevention of marine pollution from ships, Acapulco, Mexico, March 1976. IMCO, London.

11

The future

Over the past 50 years, the use of oil, initially as a source of energy, and now increasingly as a raw material for chemicals, has steadily increased, but now several things are happening. First, old resources are coming to an end and, depending upon the degree of optimism or pessimism of the prophet, the end of the dominance of oil as a source of energy and chemicals is within sight; this may well occur in the lifetime of some of us. Second, the world's primary producers have decided to have a larger share in the profits from the exploitation of their natural resources, particularly oil. Third, the dispute between Iraq and Iran has resulted in great damage to oil installations and exports have fallen severely, and, at best, it will be many months if not years before they return to the 1979 figure. This has resulted in a rapid and steep rise in the price of crude oil, and oil has changed from being a relatively cheap raw material, (even when the cost of moving it by tanker from well to refinery is included) to a valuable resource, with the prospect of further steady increases in price as the more accessible sources are exhausted. It has only been prevented from reaching even higher prices by the reduction in demand caused by the present (1980/81) world trade recession. As a result, exploration for new oilfields has moved from the land and the shallow waters onto the Continental shelves and into even deeper seas and the world is turning to areas such as the Arctic or Antarctic for supplies.

This increase in value of crude oil and the increasing international pressure should both have tended to reduce the amount of oil entering the sea. Undoubtedly tank washing, at one time the main source of pollution, has been greatly reduced and accidental spills at loading and discharging points are diminishing, but the

TABLE 11.1
MAJOR SHIPPING LOSSES (vessels over 200,000 dwt)

1969	Marpessa	206,805	Explosion
1973	Golar Patricia	216,326	Explosion
1975	Berge Istra	227,556	Explosion
1976	Olympic Bravery	277,599	Ran aground
1978	Amoco Cadiz	233,690	Ran aground
	Andros Patria	218,605	Explosion
1979	Atlas Titan	212,751	Fire
	Aegean Captain	210,257	Collided
	Atlantic Empress	292,666	Collided
	Berge Vanga	227,912	Explosion
	Energy Determination	321,186	Explosion
1980	Salem	215,000	Explosion
	Maria Aleiandra	236,000	Explosion
	Albahaab	239,410	Explosion
	Mycene	238,889	Explosion
	Energy Concentration	212,269	Broke back

last few years have seen a marked increase in the number of large tankers which have been lost. The table from the *Financial Times*, March 1981, (Table 11.1) shows those over 200,000 dwt.

The *Amoco Cadiz* was the one which produced, by far, the greatest pollution of the shore, when it suffered a breakdown in the steering gear in stormy weather. The anchor chains snapped when the anchors were dropped in an attempt to stop the tanker from drifting on to rocks. It was completely wrecked and its 220,000 tons of crude was washed up on the shores of Brittany.

Many of the other tankers shown in the table exploded during the cleaning of their tanks. This is particularly disturbing, as, after similar accidents, but to ships nearer 100,000 dwt than 200,000 in 1970, it was hoped that the causes had been found and that changed tank washing procedures would eliminate this hazard. The reasons for the present accidents probably reflect human error. The recession in the tanker industry has meant that owners have been economising in maintenance, and possibly in crew training. However, at least two of the accidents, the *Betelgeuse* and the *Energy Concentration*, have suggested an alternative. These two broke their backs while unloading apparently quite normally. Two Naval Architects, David Jones and David Adam,[1] have carried out a computer simulation of the way tankers can be loaded and unloaded with oils of different densities. They have concluded that, even in normal operating, the stresses to the tanker structure approach or even exceed the safe limit. When, to this, is added the severe corrosion which can occur, the over stressing can result in a catastrophe. This can be avoided by more careful and probably more frequent inspection of the hull and greater care in the loading and unloading procedure. The authors suggest that a simple on board computer could monitor the stresses, and give warning before the safe limit was reached.

The Eire Government Tribunal of Enquiry into the loss of the *Betelgeuse*[2] found that 'the initiating event of the disaster was the buckling of the ship's structure in way of the permanent ballast tanks; this was immediately followed by explosions in the permanent ballast tanks and the breaking of the ship's back. The ship, thereafter, was a total loss. These events were produced as a result of a conjunction of two separate factors; a seriously weakened hull due to inadequate maintenance, and an excessive stress due to incorrect ballasting on the night of the disaster . . . the absence of a loadicator contributed considerably to the error that was made'. As a result, the vessel and the jetty to which it was tied were completely wrecked and all the crew of the tanker, the wife of one member of the crew, the crew on the jetty and the ships pilot (fifty persons in all) lost their lives.

The *Amoco Cadiz* and *Betelgeuse* were, one hopes, exceptional, but as we have seen these large accidents do occur and are unlikely to stop suddenly. It is, therefore, essential to continue work to discover the best way to deal with the consequences.

Over the past 15 years a great deal of effort—both money and time—has been expended on developing new methods to deal with oil spills. These have been discussed in earlier chapters. It will have been appreciated that when oil is spilled on calm or inland waters then there are many satisfactory methods which can be used to remove it from the water. As soon as the wave amplitude and the state of the sea increases above a small minimum then, unfortunately, there is no equipment which can collect the oil.

The Editor of *Marine Pollution Bulletin*[3] in a review of a Report on mechanical systems for the removal of oil from water[4] gave it as his view; 'Unless a completely new approach is made to devising methods of picking up oil at sea, we are left with dispersants as the only effective weapon in our armoury. The only alternative is to let most of the oil come ashore. We know that in various ways it is damaging and certainly more troublesome to deal with once it gets close to shore. There is no evidence to show that it does any harm to disperse it well out to sea.'

As has been said, the major damage caused by oil pollution of the sea is the damage to the amenities increasingly enjoyed by man. The major damage to living resources is to bird life. This has been discussed in Chapter 3. Direct damage to marine life, particularly commercial fisheries, has been negligible and has only occurred in massive oil spills close to a sensitive coast, such as the *Amoco Cadiz* (but not the *Torrey Canyon*) and in highly specialised spills of toxic products in inland or confined sea areas. Permanent, but localised, damage can be caused by continuous discharges of an effluent containing low levels of petroleum products from landbased installations, such as oil refineries, chemical plants and so on. In this case, the damage is to the intertidal fauna of a restricted area, and discharges of this kind can be reduced to any desired level using existing, but perhaps costly, techniques. It has often been suggested that marine organisms would ingest oil and could concentrate it within their tissues, and it was postulated that once oil has been assimilated then it will remain within the organism for the rest of its life. On the presumption that some of these hydrocarbons are polycyclic hydrocarbons (among these are a number of well known carcinogens) this problem has been viewed very seriously by the oil industry. A large number of laboratories throughout the world have undertaken lengthy experiments to prove or disprove these hypotheses.

The American Petroleum Institute[5] has reported on a 4-year programme which involved the work of eight well known American Institutions. The Conclusions are well worth quoting:

1. Marine organisms subjected to sublethal concentrations of oil do accumulate petroleum hydrocarbons. However, these organisms rapidly lose these materials once their environment is no longer contaminated by oil.
2. There is little evidence that oil becomes concentrated by progression up a food chain. Consequently, its threat to the health of man, through ingestion of potential carcinogens, appears to be remote or even non-existent.
3. Bioassay results have limited value in assessing the probable impact of an oil spill in the marine environment.
4. If damaged at all by an oil spill, phytoplankton have the capacity to regenerate and repopulate quickly. At low levels of contamination, i.e. a hundred parts per billion or less, oil is a nutrient for phytoplankton productivity.
5. Actual experience has shown that, in most instances, an oil spill does not cause significant damage, even temporarily, to marine life. Only rarely is damage sustained that persists as much as a year.

This would seem to be conclusive. It is supported by a good

deal of similar evidence which has been produced by laboratories in other countries, particularly Western Europe. The harmful effects of oil can therefore be restricted to their visible effects; the visible contamination of beaches and boats and the death of diving sea birds. The worldwide nature of oil pollution has had one fortunate side effect. It has drawn attention to the fact that the oceans of the world are not infinite. They cannot be used as a sink or rubbish tip for any quantity of any material, whether it is harmless or harmful.

The United Nations Organisations set up a committee of scientists to examine the possible harmful effects of all varieties of marine pollution considering amenity, hazards to human health, hazards to marine life or to industrial processes. They examined many types of pollution and drew up a ranking list in which oil and oil dispersants were fifth in order of importance.[6] The list is as follows:

1. Domestic sewage, including processing wastes.
2. Pesticides.
3. Inorganic wastes, including heavy metals.
4. Radioactive materials.
5. Oil and oil dispersants.
6. Petrochemicals and organic chemicals.

The work of this Group preceded the International Conference on Marine Pollution held in 1973. This Conference culminated in the MARPOL Convention. Annex I of the Convention gives detailed rules for the prevention of oil pollution which were intended to replace the 1954–1971 Convention, while Annex II listed a large number of substances which could be harmful to the marine environment. The Convention also details the ways in which these materials should be handled on board ship to avoid pollution. Nations are invited to ratify this Convention and introduce national legislation to deal with its provisions. The Convention also listed certain special areas, the Mediterranean, the Black Sea, the Baltic, the Red Sea and the 'Gulfs' area. Here only 'clean ballast' may be discharged. Countries bordering a special area are expected to draw up a local convention or treaty to deal with all types of pollution. Because of the complexity of the Convention, and the costs which would be involved, Nations are very slow to ratify it. The prospects were improved by the Protocol of 1978 which enabled States to ratify the Convention for Oil Pollution Only. It is now hoped that it will be in force by 1982.

The enforcement and the non-infringement of all of these regulations will, in time, allow the sea to return to its pre-Industrial

Revolution state. The outlook is hopeful; more people are taking more care, not only about oil pollution but about the pollution of water by many other more harmful substances. We realise we live on one small planet. One man's waste can be another man's poison. We must all combine to preserve our fragile environment.

References

1. *New Scientist*, p. 350 (31 July 1980).
2. *Report on the Disaster at Whiddy Island, Co. Cork, Eire*, Government Publication, Dublin, 1980, para. 22.1.1, p. 321.
3. *Marine Pollution Bulletin* **6** (5), 66 (May 1975).
4. *Mechanical Systems for the Recovery of Oil Spilled on Water*, Institute of Petroleum, Applied Science, London, 1975.
5. MERTENS, E. W. and HAXBY, L. P., 'Recent advances in understanding the effects of oil in the marine environment', Panel Discussion 25(3) *World Petroleum Congress*, Tokyo, 1975.
6. IMCO/FAO/UNESCO/WHO/WMO/IAEA Joint group of experts on the scientific aspects of marine pollution, GESAMP II/11 1970.

Appendix 1

Recommended treatment of oiled sea birds

(Taken from a pamphlet written by the Research Unit on the Rehabilitation of Oiled Sea Birds, Department of Zoology, University of Newcastle-upon-Tyne, NE1 7RU and published by the Advisory Committee on Oil Pollution of the Sea.)

Essential facilities and requirements

The treatment of oiled seabirds is a skilled task and requires certain basic facilities which are unlikely to be available at improvised centres. If they cannot be provided, the rescue operation should be envisaged in two stages: first, the rescue of oiled birds from the beach and the preliminary treatment which can be carried out in improvised centres with modest facilities; second, transportation of the birds to properly equipped cleaning and rehabilitation centres.

The following are absolutely essential facilities in a cleaning centre.

(a) An abundant supply of piped hot water.
(b) A piped cold water supply with good pressure.
(c) Indoor shelter for the birds, which can be heated, particularly in cold weather.
(d) An outdoor pen with sand or concrete flooring (good turf may be acceptable providing the pen can be moved to prevent muddy or fouled conditions developing).
(e) Continuous access from the pen to a pool of clean bathing water.
(f) A reliable source of food (e.g. fresh or frozen sprats or other small fish for most seabirds). This will almost certainly

involve substantial frozen food storage facilities which are stocked when appropriate fish are locally available.

Cleaning and rehabilitation of oiled seabirds should not be attempted unless these minimal facilities can be provided, otherwise the birds will probably not be fit to release for many months, if at all.

Except for keeping rescued birds warm, members of the public should not attempt further treatment unless both experienced help and adequate facilities are available.

Capture

The sooner birds start to receive care and attention after they have been oiled the better their chance of survival. Casualties often float in shallow water without coming ashore. These can be collected from a boat or by wading. A landing net about 2 ft in diameter and mounted on a short pole will be found useful for this and can also be used for catching birds on the beach.

Always approach birds slowly and cautiously. Never chase them. Those that have already come ashore should be approached from the seaward side. The birds are sick and probably in a state of shock; handle them carefully.

Once the birds have been captured avoid frightening them by unnecessary human interference.

Preliminary treatment

Oiled birds should not be washed immediately after capture (this can be delayed for several days). The first priority is to keep the birds warm, quiet and well fed.

1. Wrap each bird in a dry cloth with only its head and feet projecting (i.e. in a poncho). This will prevent the bird struggling and exhausting itself, stop it preening and swallowing more oil than it has already done, and most important, keep it warm.
2. The birds should be transferred to a warm, draught-free place and kept in separate boxes or cardboard cartons where they can remain undisturbed. They should stand on rags or foam rubber, not on a hard surface, and these should be replaced and cleaned regularly. Do not use hay, straw or similar bedding material because it exposes the birds to the risk of contracting aspergillosis. (This is a fungal infection of the respiratory tract;

the spores of the infecting organism are harboured in mouldy hay, straw or similar material.)

3. Feeding should begin as soon as possible after the birds have been captured. Initially give strips of filleted white fish (preferable), small sprats or similar small fish, or other appropriate food. Force-feeding is usually necessary at first but it should not be attempted by the inexperienced.
4. Most oiled birds suffer from enteritis. This can be controlled by giving: Kaogel (0.5 ml once or twice a day)
   or Dianimol (0.5 ml once or twice a day)
   or Kaobiotic (a small piece, about one eighth of a tablet, twice a day)

   Kaogel is probably the best, but Kaobiotic is very effective and is easiest to administer. This treatment will probably continue for two or three days.
5. Complan can be used as a dietary supplement or replacement at first, and the birds can be given cod liver oil and multivitamin tablets, though none of these appear to be essential.
6. Seabirds' feet are often very susceptible to drying and callousing. Regularly coat the feet with a thin layer of hand cream, for example (medicated ointments such as Hibitane or Dettol ointment have been found useful).

Transportation

In many cases, oiled birds come ashore in places some distance from established cleaning and rehabilitation centres. It is obviously better if the birds do not have to be subjected to the extra strain of a long journey, but it is absolutely essential to have adequate cleaning facilities and if these are not available locally, the birds must be taken to a place where they can be properly cared for.

If transportation of birds is well managed there is no reason why it should lead to many deaths and the casualties will be far fewer than if the birds are cleaned and treated in inadequate facilities improvised on the spot.

If the cleaning centre is more than half an hour's journey from the beach, the birds should be rested and fed for at least 12–24 h before being moved. Adequately fed birds can safely be left for four or more hours without food.

The birds should be transported in warm and draught-free conditions. The birds are also most likely to remain quiet and still if they are in darkness. Wire cages should therefore not be used and the birds should be placed in individual cartons or boxes if possible. Since they generally do not preen while they are in the dark boxes, special precautions to prevent the birds ingesting oil are not necessary and there is no need to wrap them in ponchos for the journey. The boxes should be lined with rag or newspaper, particularly if the birds cannot stand.

A loosely closed box or one with small ventilation holes is usually adequate, but care must be taken not to pack the birds in boxes too tightly otherwise some birds will be in danger of suffocating.

The birds should be accompanied on long journeys by an experienced person who can examine and feed them every four hours or so.

Selection of birds for treatment

If it is necessary to limit the number of birds that are treated, the extent of oiling is not a good indication of their chance of survival. Heavily oiled birds may have come ashore quickly and therefore not be starved and not have had time to preen and swallow much oil. Lightly oiled birds may have been at sea for some days and become severely debilitated and may have swallowed a lot of oil while attempting to clean themselves.

Irrespective of the degree of oiling, birds which are thin, with a sharply protruding breastbone, or which pass large quantities of oil in their faeces have the poorest chance of survival. They should be given second priority for treatment and if resources are stretched, these are the birds that should be humanely destroyed.

If possible, delay cleaning until the birds have obviously gained in strength and weight. The birds should be alert and active and well able to stand. This may be a few days after capture and obviously depends on the initial state of the bird.

In a large oil spill with numerous casualties, pressure on accommodation and facilities may enforce earlier washing than is desirable. In this case, rest and feed the birds overnight and wash the strongest and most active birds first, leaving the others for as long as possible to recuperate before treating them.

Cleaning

The following method, if followed with meticulous attention to detail, will leave birds with a clean, water-repellent plumage, and this is essential if they are to be released quickly. The treatment involves washing the birds with a warm solution of washing-up liquid. Many detergents will remove the oil, but most have other disadvantages. Washing-up liquids are by far the most satisfactory, the cheapest liquid being the best. In UK experience, more expensive brands, whatever their merits for washing dishes, have not usually given consistent results with oiled birds and are not recommended.

7. Wash the birds thoroughly by the following method:

(a) Make up a 1% solution of washing-up liquid in a tub of hot water.
This is about:
¼ pint detergent in 3 gallons water or
100 ml detergent in 10 litres water
The solution should be hand hot (40–45°C or 104–113°F): this is the highest temperature at which the hands can comfortably be held in the water.

NOTE: It is important to have the solution within the recommended temperature range and to use the recommended concentration of washing-up liquid in water. Do not use a stronger solution or the efficiency of the washing-up liquid at removing oil will be reduced.

(b) Immerse the bird up to its neck in the cleaning solution and hold it immobile for 10 s or so, to allow the detergent to penetrate the oily feathers.

(c) Two people are usually needed to carry out the washing; one to hold the bird, the other to agitate the plumage in the solution. The oiled parts of the plumage should be rubbed and the feathers separated to allow the detergent to reach all the oil. The head and area around the eyes should be cleaned by the gentle use of a toothbrush, repeatedly dipping it in the solution.

(d) Transfer the bird to a fresh bath of detergent solution at the same concentration and temperature as before and complete washing until all visible traces of oil are removed. If a bird is exceptionally heavily oiled, a third, fresh wash, is often necessary.

(e) Most Important. The bird must be thoroughly rinsed with hand-hot water (40–45°C). A hair rinsing hose attached to a hot and cold water supply is best for this and the water should be jetted strongly into the plumage, working against the grain so that it penetrates to the skin. Continue spraying all parts of the plumage until the appearance of the bird changes from being utterly waterlogged to regaining some of its former fluffiness. The water begins to pearl off in drops. This condition should be achieved in all parts of the plumage.

NOTE: Complete removal of the oil and sufficient rinsing to remove the final traces of detergent from the plumage are essential if the birds are to regain a water-repellent plumage. The cleaning process may take some time, but unless it is thorough, the birds cannot look forward to an early release.

8. Dry the bird by mopping off excess water with a clean cloth and then place it near a source of heat. Coat the feet thinly with hand cream or antiseptic ointment. Hot air driers are not recommended because of the danger of scorching the bird and causing damage to the feet and other exposed parts. Overhead infrared heaters are best.
9. When dry, the bird should be left in a warm, draught-free place for about 3 h. It can then be transferred to an outdoor pen.

Aftercare

10. The birds should be kept in an outdoor pen with access to a pool of clean water. It is very important that the surrounds of the pond and the enclosure should be such that the birds do not become soiled or muddy. Clean sand or possibly sand on concrete is best. A good turf is acceptable provided the pen can be moved if there is a chance of muddy or foul conditions developing.

    Auks (guillemots and razorbills) will crowd together against a wall. An irregular wall of rocks 18 inches to 2 ft high near the pool appears to provide them with very favourable surroundings.
11. During the first few nights it is advisable to pen the birds under shelter where they can be given a heat source. Initially, this may also be necessary during the day in very severe weather. The birds should be hardened off, however, at the earliest opportunity.
12. If the bird has regained sufficient strength by the time it is washed, and washing and rinsing have been adequate, it will bathe frequently and preen vigorously. This is important and birds that are initially unwilling to enter the water should be

encouraged to do so. Final traces of oil and detergent are removed from the plumage by bathing and preening so that it becomes fully water-repellent within two or three days.

NOTE: It is important to provide conditions in which the birds can keep themselves clean. By frequent bathing in clean water, and preening, the birds can keep their plumage in a water-repellent state but undue contamination of the feathers with oil, detergent residues, dirt, powders or faecal matter will all cause loss of water-repellency which the bird cannot repair. If this happens it may be necessary to repeat the washing and cleaning process.

Feeding

It is important to provide appropriate food according to the species of bird and its conditions, although in an emergency the choice of food may be governed by what is available.

Auks, cormorants and shags, sea ducks and eiders, and sawbills (mergansers) can be fed on fish. At emergency centres where preliminary treatment is given, the birds should be fed strips of filleted white fish and this diet is suitable for all birds for the first day or two. The mainstay later and at the cleaning and rehabilitation centres is sprats or sand eels. Since the season for these is fairly short, adequate supplies can be assured only if stocks are frozen while they are available and held in reserve. All fish must be of good quality and very fresh or, if frozen, recently thawed. Cormorants and shags can be given larger fish such as herring or strips of cod or ling; eiders can also be given small mussels. If fresh food is not immediately available scoters and eiders can be given bread and milk.

Diving ducks, surface feeding ducks, geese and swans can be given a mixture of wheat and turkey starter pellets with added green food. Pond weeds can also be given to geese and swans, and they should be allowed to graze grass if possible. Most will take grain mixed with some form of poultry pellets with additional animal food such as chopped liver or meat.

Gulls can be fed on a wide range of animal foods including, in an emergency, tinned cat or dog food but are usually most conveniently fed with fish.

Waders can be given, worms, fresh and dried shrimp and minced animal food, such as chicken intestines, to which vitamins are added.

Release

13. As soon as the birds are strong and healthy and have a completely water-repellent plumage they can be released. This can be as little as 2–3 weeks after capture.
14. It is important to ensure that the plumage is fully and naturally water-repellent before the birds are released. The use of synthetic waxes such as Purcellin is not necessary and is not recommended because it may confer only temporary water-repellency with the result that the birds become waterlogged some time after release. The birds should be able to spend a considerable time on or in the water without wetting and they should emerge looking as dry as when they entered it.
15. Many seabirds will need to be taken to a suitable coast to be released though some ducks, given the opportunity, will leave their pens of their own volition when they are ready. Birds should be released in quiet weather and preferably sufficiently early in the day that their movements can be followed for 2–3 h after release. If they have been released prematurely they may return to shore with a wet plumage and must be collected up for further treatment.
16. It is an elementary precaution not to release cleaned and rehabilitated birds on coasts where oiled birds are still coming ashore.

Appendix 2

Dispersants approved in the United Kingdom

Oil Dispersants and Surface Film Chemicals approved by Warren Spring Laboratory and licensed for use by the Ministry of Agriculture, Fisheries and Food under the Dumping at Sea Act 1974. Intending users must apply to the appropriate Authority for a licence to use approved dispersants. A number of other countries have similar prohibitions. *June 1980*

A. Hydrocarbon Solvent Based Dispersants

1. DASIC SLICKGONE LT2 Dasic International Limited Winchester Road, Romsey, Hants, SO5 8YD. Tel: Romsey 512419.	Sea and Beach use
2. COREXIT 8354 Essochem Division de SA Esso NV Mechelseteenweg 363 B-1950 Kraainem Belgium. Telex: 22364	Sea use only
3. EMULSOL LW Arrow Chemicals Limited PO Box No. 3 Stanhope Road Swadlincote, Staffs. Tel: Burton-on-Trent 221044	Sea and Beach use
4. FINASOL OSR-2	Sea use only

5. FINASOL OSR-3 Petrofina (UK) Ltd., Petrofina House York Road, London SE1 7NT Tel: 01–928–8000	Sea use only
6. LANKROMUL OSD Lankro Chemicals Group PO Box I Eccles, Manchester, M30 0BH. Tel: 061–789–5311.	Sea and Beach use
7. SHELL DISPERSANT LTX Shell Chemicals UK Ltd., Villiers House, 41/47 Strand London WC2 Tel: 01–438–3678.	Sea and Beach use
8. GAMLEN OSR LT 126 Gamlen Chemical Co (UK) Ltd., Wallingford Road, Uxbridge, Middlesex, UB8 2TD. Tel: Uxbridge 32412	Sea use only
9. AMEROID OIL SPILL DISPERSANT/LT Drew Ameroid UK Ltd Marlowe House Station Road, Sidcup, Kent. DA15 7EP. Tel: 01–300–0102	Sea and Beach use
10. FLEETEX BD/3 Isaac Bentley & Co Ltd Naylor Street, Liverpool, L3 6DS Tel: 05–227–3441.	Sea and beach use
11. KRAKEN MC 563 The Kraken Group Cwmdu Industrial Estate Swansea, SA5 8JK. Tel: Swansea 33686.	Sea and beach use
12. EMKEN SPILLWASH LT Emken International Limited The Point,	Sea and beach use

Leigh Beck,
Canvey Island, Essex,
SS7 SH2.
Tel: Canvey Island 62161.

13. BMD (SM 53) — Sea and beach use
Smyth-Morris (Chemicals) Ltd.,
513 Bradford Road,
Batley, Yorks. WF17 8LW.
Tel: Batley 477201.

14. SOS — Sea and beach use
Safeform Oil Slick Dispersant

Momar Industrial Services Ltd.,
Mercian House
York Street,
Sturminster-on-Severn,
Worcs., DY13 9EE.
Tel: Sturminster 6232.

15. VECOM B14-25 GL — Sea and beach use
Vecom Ltd.,
Link House,
56/58 High Street,
Billericay, Essex, CM12 9BS.
Tel: Billericay 55577.

16. OILSOLVE — Sea and beach use
Antox
Steetley Chemicals Ltd.,
West Street, Swadlincote,
Nr. Burton-on-Trent,
Staffs. DE11 9DG
Tel: 0283 215659

17. BP 1100X — Sea and beach use
BP Oil Ltd., Detergents Divn.
BP House,
Victoria Street,
London SW1E 5NJ
Tel: 01–821–2652

18. PERO-KLEAN 868 — Sea and beach use
The Perolin Company Ltd.,
Brimscombe Mills,
Brimscombe,
Stroud, Gloucestershire, 9LS
2SA.
Tel: Brimscombe 2581

19. ARDROX SR 61 Ardrox Ltd., Furlong Road, Bourne End, Bucks., SR 61. Tel: Bourne End 24951	Sea and beach use
20. APPLIED 8-40 Applied Chemicals Ltd., Salisbury Road, Uxbridge, Middlesex. UD8 2SW. Tel: Uxbridge 37151.	Sea use only
21. SOLVOIL Custodian Industrial Chemicals Ltd., Bridge Street, Northwich, Cheshire. CW9 7NR. Tel: 0606 44661	Sea and beach use
22. AQUARITE AWX Albright & Wilson Ltd., PO Box 80, Trinity Street, Oldbury, Warley, West Midlands, B69 4LN. Tel: 021 552 3333.	Sea and beach use
23. EBBCLEAN Bestobell Chemical Products Ltd., 131, Western Road, Mitcham, Surrey. Tel: 01–648–3422.	Sea and beach use
24. ROCHEM OIL SPILL REMOVER (WSA) Rochem International Ltd., 1, High Street, Ruislip, Middlesex, HA4 7AU. Tel: Ruislip 31381.	Sea and beach use
25. ATLANTOL 3211/E Atlantol Laboratory PVBA ECO SPRL	Sea use only

	Meersweg 119 9810 Drongen, Belgium. Telex ECOTOL 11982	
26.	ACMA EP 540 Agma Chemical Co. Ltd., The Leazes, Hexham, Northumberland NE46 3RW. Tel: Hexham (0434) 4611/4.	Sea and beach use
27.	X-3125 M. W. Webber & Co. Ltd., Springfield Road Oil Works, Highbridge, Somerset, TA9 3AB Tel: Burnham-on-Sea 3234/5/6	Sea and beach use
28.	PETROCON OIL SPILL ELIMINATOR IV Petrocon Marine Chemical Corp Two World Trade Centre, New York NY 10048. Telex 222 804.	Sea use only
29.	SERVO CD 2000 Servo Chemische Fabriek, B. V. Delden, Holland, Telex 44347	Sea and beach use
30.	SCOTRAM LTOD 2000 R. S. Guthrie & Co (Scotland) Ltd., 9a West Maitland Street, Edinburgh EH12 5DU Tel: 051 229 6755	Sea and beach use
31.	APC LTOD HG Anti-Pollution Chemicals Ltd., Wardpark South Industrial Estate, Cumbernauld,	Sea and beach use

G67 3HS. Tel: Cumbernauld 31081.	
32. OSD 88 R. P. Adams, Arpal Works, Riverside Road, Selkirk, TD7 5DU. Tel: Selkirk 21586	Sea and beach use
33. TEKLENE TC 48 Tekchem Ltd., Erimus House, Queen's Square, Middlesbrough, Cleveland. TS2 1AA Tel: (0642) 221441	Sea and beach use
34. AQUACERF 25 Cerfact Europa B V P Box 208 Haarlem Holland	Sea and beach use
35. KEMARINE 66 Kem Europa B V P Box 199 Haarlem, Holland.	Sea and beach use
36. EMulso E-309 Toho Chemical Industry Co., Ltd., No. 14–19, I-Chome, Nihon Bashi, Kakigara-Cho Chuo – Ku Tokyo, 103 Japan. Telex NBR 2522 332 J	Sea and beach use
37. OD II Gremer Chemicals UK Ltd., Progress Works, Tywern Road, Rhiwbinia, Cardiff Tel: 0222 61 6917	Sea and beach use

38. Remospill I Performance Chemicals 60 Verney Road, London SE16 3DH. Tel: 01-231-3737	Sea and beach use
39. K559 Burmah-Castrol Ltd., Burmah House, Pipers Way, Swindon SN3 1RE Tel: 0793-30151	Sea use only
40. Techmar 125 Spillclean Techmar Technical Consultants & Chemical Suppliers, 7 Tudor Road, Southport, Lancs TR8 2RY Tel: 0704–75759	Sea and beach use
41. Slicksol Oil Dispersant Larragard Ltd., 3 Manor Drive, Halifax HX3 0DN, West Yorkshire.	
42. Applied 7273 Applied Chemicals Ltd., PO Box 43, Salisbury Road, Uxbridge, Middlesex UB8 2SW	

Appendix 3

The WSL dispersant spraying equipment may be obtained from:

United Kingdom	Biggs Wall Fabricators Ltd., Hampden House, Arlesey, Bedfordshire, England SG15 6RT Tel: 0462 73133 Telex: 826113
All Other Areas	Vikoma International Ltd., Littleton Road, Ashford, Middlesex TW15 1UT England. Tel: 07842 45011 Telex: 887767 (VIKINT G)

Appendix 4

The dispersant spraying equipment may be obtained from:

Frank Ayles & Associates Ltd,
120 Whitechapel High Street,
London E1 7PT
England.

Tel: 01 247 1926

Telex: 886089

Harrier Marine Ltd,
Pilsworth Rd,
Bury,
Lancashire

Tel: 061 796 8703

Telex: 666754 Harpex 6

List of figures

List of tables

Acknowledgements

Fig. 1.1 Shell International Petroleum Co.
Figs 1.2, 1.3–1.5, Table 1.2 International Tanker Owners Pollution Federation
Fig. 1.3 RNAS Culdrose (Crown copyright)
Fig. 1.4 J. Wardley-Smith
Fig. 1.6 P. G. Jeffery (Laboratory of the Government Chemist)
Table 2.1 H. J. Marcinowski
Fig. 2.1 P. G. Jeffery (Laboratory of the Government Chemist)
Fig. 2.2 J. Wardley-Smith
Table 2.3 Ministry of Defence Research Information Centre
Fig. 2.3 F. Smith
Fig. 4.1 A. Nelson-Smith (Paul Elek, publishers)
Figs 4.2–4.7 Jenifer M. Baker
Fig. 4.8 'Fate of oil at sea' (Ministry of Defence)
Fig. 4.9 The Royal Society for the Protection of Birds and Dennis Green (photographer)
Figs 5.1–5.8, 5.14, 5.16–5.20, 5.22–5.28 D. E. Newman (Late of the Hydraulics Research Station, Wallingford)
Figs 5.12, 5.13 BP (Sunbury Research Centre)
Fig. 5.15 William Warne & Co.
Fig. 5.21 Hydraulics Research Station (Crown copyright)
Fig. 5.29 HMS Daedalus, HM Dockyard, Portsmouth
Figs 6.1, 6.2 P. G. Jeffery (Laboratory of the Government Chemist)
Figs 6.3–6.6 BP
Fig. 6.8 Oil Mop Inc.
Figs 6.10, 6.12–6.15 Warren Spring Laboratory
Fig. 6.16 Shell Development Co, Houston
Table 7.1 Fuller, 'Journal of the Institute of Petroleum'
Fig. 7.1 J. Wardley-Smith
Figs 7.2, 7.3 Rocket Propulsion Establishment
Table 8.1, Figs 8.1, 8.7 J. Wardley-Smith
Fig. 8.2 Warren Spring Laboratory
Figs 8.3, 8.4, 8.6, 8.9, 8.10 R. J. Farn (BP)
Fig. 8.5 MAFF
Figs 8.8, 8.11 BP
Table 8.2 D. Cormack
Table 8.3 J. A. Nichols and I. C. White
Figs 9.1, 9.5, 9.7–9.9 Warren Spring Laboratory
Fig. 9.2 Massey-Ferguson
Fig. 9.3 PACE
Figs 9.4, 9.10 J. Wardley-Smith
Fig. 9.6 Cooper-Pegler
Fig. 9.11 Institute of Petroleum
Fig. 9.12 Wessex Water Authority
Table 10.1, Figs 10.1–10.9 J. Wardley-Smith

Subject index